THE FLIPSIDE

THE FLIPSIDE

*How to Invert Your Perspective and Turn
Fear into Your Superpower*

Michelle "MACE" Curran

New York   Boston

The views expressed in this publication are those of the author and do not necessarily reflect the official policy or position of the Department of Defense or the U.S. government.

The public release clearance of this publication by the Department of Defense does not imply Department of Defense endorsement or factual accuracy of the material.

Copyright © 2025 Michelle Curran

Cover design by Jim Datz
Cover photograph (pilot) © Michelle Curran
Cover photograph (clouds) © Shutterstock
Cover copyright © 2025 by Hachette Book Group, Inc.

Hachette Book Group supports the right to free expression and the value of copyright. The purpose of copyright is to encourage writers and artists to produce the creative works that enrich our culture.

The scanning, uploading, and distribution of this book without permission is a theft of the author's intellectual property. If you would like permission to use material from the book (other than for review purposes), please contact permissions@hbgusa.com. Thank you for your support of the author's rights.

Grand Central Publishing
Hachette Book Group
1290 Avenue of the Americas, New York, NY 10104
grandcentralpublishing.com
@grandcentralpub

First Edition: September 2025

Grand Central Publishing is a division of Hachette Book Group, Inc. The Grand Central Publishing name and logo is a registered trademark of Hachette Book Group, Inc.

The publisher is not responsible for websites (or their content) that are not owned by the publisher.

The Hachette Speakers Bureau provides a wide range of authors for speaking events. To find out more, go to hachettespeakersbureau.com or email HachetteSpeakers@hbgusa.com.

Grand Central Publishing books may be purchased in bulk for business, educational, or promotional use. For information, please contact your local bookseller or the Hachette Book Group Special Markets Department at special.markets@hbgusa.com.

Print book interior design by Taylor Navis

Library of Congress Cataloging-in-Publication Data

Names: Curran, Michelle, author
Title: The flipside : how to invert your perspective and turn fear into your superpower / Michelle "MACE" Curran.
Other titles: How to invert your perspective and turn fear into your superpower
Description: First edition. | New York : GCP, 2025.
Identifiers: LCCN 2025012952 | ISBN 9781538768105 hardcover | ISBN 9781538768129 ebook
Subjects: LCSH: Curran, Michelle, 1987- | United States. Air Force.
 Thunderbirds—Biography | Women air pilots—United States—Biography |
 Fighter pilots—United States—Biography | Stunt flying—United States |
 F-16 (Jet fighter plane) | Self-confidence
Classification: LCC UG626.2.C87 A3 2025 | DDC 358.40092
 [B]—dc23/eng/20250518
LC record available at https://lccn.loc.gov/2025012952

ISBN: 9781538768105 (hardcover), 9781538768129 (ebook)

Printed in Canada

MRQ-T

1   2025

To all of you who have joined me in chasing big dreams with courage and boldness. I hope you see yourself in these pages and that this book inspires and challenges you.

CONTENTS

Introduction: The Surprising Flipside of Fear ix

Chapter One: Go to War with Your Inner Critic 1

Chapter Two: Everyone Has a Call Sign 24

Chapter Three: Stop Playing Dress-Up 44

Chapter Four: Remember to Wiggle Your Toes 71

Chapter Five: Take One Minute, One Hour,
One Month for Change 88

Chapter Six: Don't Get Stuck in the
Comfortable Misery 107

Chapter Seven: Seek Silver Linings 125

Chapter Eight: Focus on the Next Closest Alligator 143

Chapter Nine: You're Better Lucky Than Good 162

Chapter Ten: Avoid Get Home–itis 180

Chapter Eleven: Check Your Six 197

Chapter Twelve: You're Saying There's a Chance 215

Acknowledgments *235*
Endnotes *239*

INTRODUCTION
The Surprising Flipside of Fear

I can fly an F-16 fighter jet upside down, but I'm afraid of spiders. Put me in a cockpit 150 feet above the ground and tell me to fly straight up toward the clouds while giving the jet a few spins, and I'm your girl. Point at something no bigger than my thumb, its eight legs making their creepy way toward me, and I'm out the door faster than my sheepadoodle when he spots the Amazon truck.

I know that probably doesn't square with your image of a fighter pilot. I'm supposed to uphold the public image of our military: willing to fly, full of swagger, into combat zones at a moment's notice, preferably in cool aviator sunglasses. Brave. Unflinching. Fearless. Yet throughout my entire life, through my flight training, and during my early days in the cockpit, I found myself battling an unexpected and relentless enemy: my inner critic. It's led to a lifelong battle with imposter syndrome, when you doubt your own skills, talents, and even accomplishments. It may seem odd that someone who has forged a career performing aerobatic maneuvers inches from another jet would doubt her own abilities, but it's true.

INTRODUCTION

I'd made it to the Air Force's elite air demonstration squadron, the Thunderbirds, as just the fifth woman to ever perform aerial maneuvers in its seventy-year history. I was the Lead Solo—the person who takes on the highest gravitational force of any position on the team—spending a great deal of time inverted. Throughout my career as a fighter pilot, I endured claustrophobia, do-or-die situations, and the effects of nine times the force of gravity on my body, but my biggest adversary became my own voice of doubt.

No matter what I achieved in school or in the military, or how many times I flew my F-16 upside down over a delighted crowd at Thunderbird shows, a nagging doubt coming from within kept me from fully seeing who I could become. I was driven enough to keep trying, but I'd get in my own way, stuck in my head, worrying about what other people thought of me and dwelling on my failures. Speaking up in front of my peers and commanders was more difficult than any physical challenge. Being vulnerable enough to accept help when I needed it or bold enough to ask for what I wanted were lessons I had to repeat over and over before I finally learned from them. I had to figure out for myself how to override my inner critic and trust my own abilities, whether I was in the cockpit or in a debrief on the ground.

On my way to fulfill my dream to fly fighter jets, I learned that my choices between discomfort and security came with a price. I was far less scared of flying five hundred miles per hour and closer to another jet than a pitcher's mound is to home plate than I was of simply being myself because the former was more predictable. Eventually, seeing the flipside of fear would become my new guide for making smart decisions in my life. But I'd need to make a few mistakes first.

INTRODUCTION

After a potentially deadly error in the cockpit—the one that gave me my call sign—my doubts about whether I was good enough for the career I was working so hard to achieve became more intense than ever before. My inner critic stepped in to talk nonsense in my head, and I believed it.

I was at a particular disadvantage because of my own disposition, which didn't always fit in with the type A fighter pilot culture of cocky risk-takers behaving like Tom Cruise's Maverick in the *Top Gun* movies. I was the introvert too embarrassed to raise my hand during training and later, uncomfortably signed autographs for excited fans after a Thunderbirds air show. Though I enjoyed inspiring people, it was easier for me to do that while flying my jet belly to belly with another at four hundred miles per hour as I shot by overhead.

Along the way, I caved in to my own fear more times than I could count, avoiding opportunities or staying silent when I could have spoken up. Yet I eventually saw that avoiding discomfort came at a higher cost than facing it head-on because it forces us back into our comfort zones, which stunts our growth and keeps us emotionally, spiritually, and personally small. And that's no way to chase your dreams.

To evolve from zero flight hours to Air Force pilot to F-16 combat fighter pilot to Thunderbird, I had to learn to accept what's out of my control, relying on facts instead of emotions and seeing opportunities where there are obstacles. It's not that I suddenly became fearless; I still need fear to keep me safe—we all do. It's that I changed my relationship with fear. Ultimately, I learned how to make the uncomfortable choice by flipping my perspective, like my jet, upside down.

One of the most famous scenes in the original *Top Gun* movie is

INTRODUCTION

when an instructor, played by Kelly McGillis, asks how Maverick could see the enemy jet if he was flying directly above him.

"Because I was inverted."

This has become such an iconic line that people around the world recognize it, and it has stayed in the zeitgeist ever since. There's a fascination with the idea of flying a jet upside down at insane speeds. It feels uncomfortable, brave even—and it challenges our perspective.

This is how I hope you'll begin to see things—from the flipside. I'm on a mission to teach people all I know about facing fears and making hard choices so you, too, can see things in a new way for a richer, more fulfilling life. I'll share each of the lessons I've learned through a series of fighter pilot experiences and personal principles so that you can begin to apply them to your own challenges.

Whether you've set your sights on an audacious goal or you're just trying to get through this week, looking at things from the flipside can smooth the ride or send you on a completely unexpected and incredible trajectory. I'll show you how. Let's go!

THE FLIPSIDE

CHAPTER ONE

Go to War with Your Inner Critic

People often assume that I've always wanted to be a fighter pilot, but I didn't spend my childhood pinning Thunderbird posters onto my bedroom wall or dressing up as Amelia Earhart for Halloween. The truth is, I stumbled upon this career while standing on a tarmac in Florida. I wanted to be an FBI agent, so I majored in criminal justice at the University of St. Thomas. To pay for college, I had a scholarship from the Air Force through the Reserve Officers' Training Corps, or ROTC. I figured I'd serve a few years in the military to fulfill my scholarship and then head to Washington, DC, to become a federal agent. But then I saw the jets.

By that point, I had already made it through more than two years of the ROTC program and attended the officer equivalent of basic training, ROTC Field Training, which was designed to weed out anyone who lacked the leadership potential or the physical fitness to commission as officers. I knew I was physically fit, but I wasn't sure if I had what it took to be a leader. I had an image in my head of what a military leader was, and that image wasn't me:

a shy girl from a small town who barely knew anything about the career she was headed toward.

Throughout the program, the Air Force officers in charge of the cadet wing would occasionally ask, "Who here wants to be a pilot?" I was unsure yet intrigued. It sounded exciting. I imagined flying at high speeds, spinning through the air, and tackling the toughest missions in the Air Force. No matter how much I doubted whether I could fly jets, imagining myself doing it stuck in my mind whenever the instructors brought up flying. But I also knew that if I raised my hand to show I wanted to be a pilot, everyone would look at me, and then they'd watch me throughout training to see if I succeeded or failed. I didn't want that accountability, so I didn't raise my hand—until the day we visited Tyndall Air Force Base on the Florida panhandle, and I got to see two fighter jets up close for the first time.

They weren't just sitting there on the ramp, either. They were taking off for a training flight in full afterburner. It was dusk, so the orange flames firing from the F-15s' engines were visible against the darkening sky, and I got goose bumps. The jet engines' noise, like a thunderclap, vibrated through my entire body. It was such a visceral experience that the hair on the back of my neck stood up, and I thought, *I'm going to do that.*

The next time my ROTC instructor asked, "Who here wants to be a pilot?" I raised my hand. After years of keeping my head down, I seemed to catch him off guard.

"Oh, what do you want to fly?"

In front of my entire class, I replied, "Fighters." I think I sounded confident, even if I didn't feel it.

"You know how you can tell there's a fighter pilot in the room?" he asked.

THE FLIPSIDE

"How?"

"They'll tell you."

It felt like I'd forced myself into a "fake it till you make it" situation. I'd just committed myself to attempting to become a fighter pilot, and very publicly, and I didn't even know if I was good enough to do it. I did know that the odds were not in my favor. I would have to compete against the other ROTC cadets in my wing for a handful of pilot slots, and that was before pilot training and the initial screening for pilot aptitude.

The learning curve was steep. Not only did potential pilots have to memorize new concepts quickly, but they also had to perform well on written exams, deal with recalling information under pressure, and compartmentalize the things that would inevitably go wrong. That would be followed by a year of learning to fly multiple aircraft, all in an effort to secure the one to three fighter aircraft that, historically, were available for each class. All of that had to go well, exceptionally well, if I was even going to get the chance to try my hand at flying a fighter. If I didn't meet the standards at any point in these various programs, I would be washed out and reclassified into a new career based on the needs of the Air Force. And my dreams? They would no longer matter. That's part of the risk you take when you agree to serve your country. Yet I'd gone and mustered the courage to say what I wanted out loud. Once the conversation moved on, I felt like, "Oh man. Now everyone *knows*." There was no going back.

Even though I was terrified, I was still excited enough to start imagining myself flying the F-16, my dream jet, while pushing my body through morning physical training, my lungs burning and the copper taste of blood coating my mouth. The hard part of training continued to be raising my hand to volunteer for leadership opportunities, which always led to a queasy drop in my stomach.

"Who's going to lead the formation?" my instructor would ask, looking for a volunteer to command the group of cadets for marching drills.

Drop.

I'd seen others do it and fail spectacularly in front of everyone else, marching their formations into other groups or into buildings, unable to turn them back around. It's funny to watch but embarrassing to endure when you're the one getting yelled at by the instructors. Marching large formations was a skill that'd be rarely used once you were on active duty, but it served as a great litmus test to gauge your ability to think ahead and perform under pressure.

I knew that the Air Force wasn't just looking for physically fit cadets, but that had become my strong point. I'd made sure of it. When I applied for my ROTC scholarship, I had to run a mile-and-a-half physical fitness test, but I wasn't a runner. Back in high school, when I ran it while the track coach timed me, my lungs burned and I hated every minute of it—all thirteen and a half of them. When I crossed the finish line, I thought, *I never want to do that again.*

But if I got the ROTC scholarship, that would only be the first of many more runs and at much farther distances. By midway through college, I was maxing out the test with a 100 percent score, which meant I was running the mile and a half in under eleven minutes. Eventually, I'd run it in under ten minutes, beating many of the guys, and then I didn't hate it so much anymore. It became a benchmark where I could compete against my previous times and prove myself as capable to those around me.

THE FLIPSIDE

Over the course of the ROTC program, I came to realize how much you can change and evolve and grow when you're willing to push yourself to do the uncomfortable things, no matter the context. My classmates and I were being evaluated against one another and yet, my hand, like my stomach, was usually down. I knew that raising my hand would set me apart, but my inner critic wouldn't allow it, spewing put-downs like, "You'll surely make a fool of yourself," or "Everyone will be watching if you fail." If I wanted to become an Air Force fighter pilot, I'd have to go to war with this voice of doubt, and I'd have to win.

DEFINING THE INNER CRITIC

You know how when kids hear something in the dark, they ask an adult to turn on the light and look for the scary monster they're sure is in the room? The "monster" so often turns out to be a pile of laundry or a stuffed animal staring from the corner. It's something they really didn't have to be scared of at all, and yet, the fear felt very real to them until the light was switched on, and then they could be safely tucked back into bed. For many of us, that scary thing lurking in the dark is the negative thoughts our inner critic spews.

We all have an inner critic. It is that voice that questions your choices and plans. It judges everything from the outfit you're wearing to your intelligence. Sometimes, it talks to us, and it can be pretty abusive:

"You're an idiot."
"What's wrong with you?"

"You look fat. No, you *are* fat."
"Nobody loves you."
"Who'd want you as their wingman?"

It doesn't give *constructive* criticism, but just plain criticism. You start to believe what it says is factual, even when it's not. It magnifies the bad and brushes off the good. It's not reality and it's not you.

It may act like a schoolyard bully, but it has a purpose: It checks for threats to your safety. Though it can sound pretty mean, it's designed to prevent social rejection, and, as psychotherapist Kathy Steele told the *Washington Post*: "It's trying to keep us alive."[1] It attempts to correct your behavior so that you aren't shunned from your community, which was a survival must when we were hunter-gatherers dodging hungry wild animals and avoiding hostile tribes.

Each of our inner critics is different, fueled by messages we internalized as kids. Maybe your sister got better grades, making you the "dumb one" by comparison. Or perhaps you struck out your first time at bat in Little League, and someone told you that you weren't cut out for sports. Or maybe your big brother made fun of your height, and you began to believe that you're a loping Frankenstein like he announced every time you walked into the house.

Once I joined the Thunderbirds, the inner critic served an important role because I was flying a jet just inches from my fellow pilots, sometimes just one hundred feet off the ground at five hundred miles per hour. I needed to be socially accepted as part of that team. I needed to be trusted by the pilot next to me. But that critical inner voice can also be detrimental, especially when it's spewing mean things while you're not in a life-threatening environment but just trying to get through your day.

THE FIVE TYPES OF INNER CRITICS

From my perspective, there are several categories of inner critics, all of which come to play a role in our lives. While they're rooted in safety, these criticisms still often result in doubt and anxiety. If we can identify what these critics look like, we can work to combat the unhelpful feelings they dredge up.

The Fear Critic: It says, "If you do this, you will die!" Its goal is to keep you in your comfort zone by scaring you. When you don't stretch yourself, you don't grow, and the Fear Critic wants to stunt you to keep you safe. When you're flying formation just feet from the other jets, the Fear Critic might shout, "We shouldn't be doing this!" It's that uncontrollable urge to fly away from the danger—and right out of position.

The Comfort Critic: This critic encourages you to choose the path of least resistance, the warm covers of your bed over getting dressed for the gym. It says, "This will be hard and you need time to rest." It's easy to listen to the Comfort Critic because our society is set up to support it. We Uber Eats dinner right to our doors, binge-watch entire seasons of shows in a weekend, and order that book or sneakers or bathrobe that Amazon will deliver tomorrow afternoon. Doing so keeps us occupying our time with things that feel easy and familiar rather than seeking out new experiences or pushing ourselves to grow.

The Friend Critic: Think of this critic as your best friend in sixth grade trying to keep you from making a fool of yourself in the school cafeteria. It warns, "If you do this, people might not like you." It's the voice that weighs in when you consider asking the "mean girls" if you can sit with them. The one that has you caving to peer pressure rather than going against the crowd. It keeps you liked by those around you because it makes you *conform*. When you embrace the Friend Critic, you shrink yourself in order to fit in, missing out on pursuing what makes you, you.

The Reputation Critic: This is your sixth-grade best friend all grown up and warning you not to volunteer for the big presentation at work. It says, "If you try this, people will judge you." It wants you to blend in and lay low so people don't talk about you. Social media amps this critic up a thousandfold. Where we used to worry just about the people we saw day to day, now we have dozens, hundreds, and even thousands of people watching and judging through our screens, and all it takes is one thumbs-down in a sea of thumbs-up to kick it in. The Reputation Critic keeps you playing it safe, only doing things that will be acceptable to "them."

The Wrong Critic: This critic sees only in black-and-white. It's afraid of pushback and it worries that those around us know more about any given topic than we do, and no matter how humble we might be, most of us still don't like to be flat-out wrong. This can look like rejection, or it can show up as a fear of speaking our opinion if we aren't 1,000 percent sure of the facts, unable to support and cite them like we're defending a thesis. The Wrong Critic is the one that makes us keep our mouths shut.

The Five Inner Critics can affect anyone, anytime, but they can be particularly loud if you stand out in any way. When I raised my hand and confessed that I wanted to be a fighter pilot, I was one of just a handful of women in the room, and no wonder: Women make up just 20 percent of the Air Force officer corps.[2] Female fighter pilots are even more rare. So I didn't look like the average fighter pilot: overwhelmingly male and white. As a result, whenever I did something that made me stand out, there would be extra scrutiny, and I knew it. But that day, this girl raised her hand.

Since then, I've learned how to shine a light on my inner critic. Taking a hard look at the voice in your head helps you understand why your inner critic does what it does. If you recognize the Five Inner Critics, you can begin to see them for what they are, but sometimes it's helpful to give your specific critic even more of an identity. Ask yourself what your inner critic looks and sounds like, and which situations get it riled up. Once you realize that your inner critic takes on the role of an overbearing mother, a mean teacher, or a societal expectation of the perfect mom, spouse, or leader, you can see it as something separate from yourself—something you can choose to ignore.

I realized that for much of my career my inner critic looked like an older, more experienced, male fighter pilot—basically what society thought the ideal fighter pilot should look and act like. It took years, but taking small steps outside my comfort zone has helped prove to my faceless inner critic, and to myself, that I do deserve a place at the table and the successes that have come with it.

Once you identify your inner critic, you've got to go to war with it to take away its power. This may sound dramatic, but it's an intentional look at our strengths, weaknesses, and support systems to help us build defenses against that negative inner voice. The first step is to identify your loadout.

WHAT'S YOUR LOADOUT?

If you want to take back control, you're going to have to go to battle, and that means you'll need to assess your loadout. In the Air Force, the loadout consists of the weapons on your aircraft that you can use to defeat the enemy. When you're fighting any of the Five Inner Critics, your loadout is what makes you feel confident and like yourself. It's your strengths—and the strengths worth building up.

You may not spend as much time thinking about what you're good at as you do your weaknesses, and that's by design. Humans are hardwired for negativity bias, paying more attention to negative events, like a snide comment or a thumbs-down, and making decisions based on negative data.[3] This made sense back when negative threats were a matter of life and death, and you had to pay close attention to others' admonitions or else become lunch for a pack of hyenas. But Facebook comments aren't sticks and stones, and you don't have to give them so much attention and weight.

It takes effort to focus on your strengths, but when you do, you start to shore yourself up against your inner critic. Start by taking inventory of your loadout, including your strengths at work, in relationships, and in life. Are you a good listener? Are you a good problem solver? Do you get along with just about anyone? Can you parallel-park in one smooth motion or make a tasty lasagna? These are all strengths, and you may dismiss them because they come easily to you. But for every strength you downplay, someone else wishes they had it. What are yours?

Consider how animals use their inherent strengths to survive in the wild, like the bald eagle vs. the crow. Crows have been known to attack bald eagles in flight, either to steal their food or to thwart them from nesting areas. They've even been seen riding on eagles' backs. An eagle could kill a crow, but it usually doesn't. Why? Because it doesn't want to get injured, which would hinder its ability to hunt. Without food, it would die. So instead, it uses its loadout: its ability to fly higher than the crow can endure. The eagle doesn't fight back at all, keeping its sharp talons poised as landing gear. Instead, it just slowly flies higher and higher until the crow, lacking enough oxygen, falls off its back. That's pretty badass.

If the eagle had an inner critic, though, it would probably say, "You're a bald freakin' eagle! Fight that little crow. What's wrong with you?" But birds don't have inner critics, and eagles know their own loadouts. Be like the eagle. Don't waste your time and energy on fights you don't need to have.

Identifying your passions is just as important as knowing your strengths because they provide a sense of purpose that you'll likely dig deep to fulfill. These are things you love to do that you may not even be good at yet, and still they energize you. Maybe you're learning to surf or play the piano and you're not quite proficient, but you're having fun trying. Perhaps you're the barbecue boss of the neighborhood or the karaoke queen of the local bar. We tend to underappreciate passions, relegating them to nice things to get to if we have free time on the weekend.

But because of that sense of purpose, passions can put you in a state of "flow," that feeling that you're "in the zone." Mihaly Csikszentmihalyi, a cofounder of positive psychology and the author of *Flow: The Psychology of Optimal Experience*, said, "A person can make himself happy, or miserable, regardless of what is actually happening outside, just by changing the contents of consciousness."[4] In a state of flow, he said, nothing else seems to matter, and in my experience, flow makes no room for the inner critic.

After I fell into running as a way to get in shape for the Air Force, my body adapted and running soon became a passion. If you're an endurance athlete you know exactly what I'm about to say about feeling in flow while running, but all of us have struggled through a run or a hike or other physical exertion that left us huffing and puffing and wishing it was over. Yet if you can keep showing up through that uncomfortable stage of adaptation, there's a reward on the other side—flow, and maybe even passion.

I didn't find this out until I was training for my second marathon. For the first race, I had done the work, but it felt like an uncomfortable battle the whole time. I had a lot of doubt about whether I could run that far, and my inner critic regularly filled my headspace while I slogged away one mile after another. My flow had flown away.

During the second marathon, though, I really hit my stride. Yes, literally. I would do my best runs without music or the distraction of a podcast, focusing instead on the sound of my breathing and my feet hitting the sidewalk or crunching the gravel until it lulled me into a state where the miles flew by. It was as close to time traveling as I imagine I'll ever experience. My mind's tolerance for the discomfort of running increased as my body adapted. What would have felt like forever in the past became bearable, and eventually enjoyable.

Those of you who hate running may have left the chat. This is just my example. I've also experienced this in the cockpit, on the side of the mountain I was hiking, and while writing this book. Finding the thing that takes you to this level of focus and flow,

THE FLIPSIDE

the thing that silences your inner critic as it becomes part of your loadout, is highly personal. Once you find this place of solace from self-doubt and from day-to-day stressors, you can return to recharge and gain perspective when you need it.

LIST YOUR DRAWS

If your loadout is made up of your strengths and passions, your draws are your next-level hopes and dreams, the things you've felt drawn to, even called to do, but you've been too scared to go after. Maybe you've always wanted to write a screenplay, hike the Pacific Crest Trail, start a nonprofit, or turn your beloved brownie recipe into a sweets business. These are draws, and they're important because they can motivate you to get unstuck, try new things, and grow. They may even lead you to a new career, like seeing those jets did for me.

But something may be keeping you from even trying. You may think you're not good enough to give it a go. You think of someone else who's doing your draw and doing it well and think, *I can't do it like they do, so why bother?* But they didn't start out that good at your draw, and, as I'll go into detail in chapter three, we're all beginners at something. The more you learn about your draw, the more experienced you get. You just have to take that first step. I mean, I wasn't born a fighter pilot!

You may think you're not ready or that you don't deserve your draw, as though you're not worthy of the happiness that comes with it. You believe that's for the special few: the rock stars, soccer MVPs, *Shark Tank* investors, senators, Instagram influencers.

Fighter pilots, maybe? Then how come your draw still calls to you? Why do you still daydream about it?

When you write down your draw, you make it real. Then it's no longer a wild fantasy you indulge in when you're having a bad day or staring out a window. Take your draw off your bucket list and put it on your to-do list. You could start small, maybe with a book to read or a podcast to listen to, or you could simply say your draw out loud. Then get ready for your inner critic to crap on it.

Draws often are rewarding. Hard, but rewarding. I started to see a pattern: Those opportunities that piqued my interest but also made me feel a little bit scared were the most gratifying. It's like Eleanor Roosevelt reportedly said: "You must do the thing you think you cannot do." It was that lesson of going for scary things and making bold decisions that led me to even consider applying to the Thunderbirds—one of the biggest draws of my life.

The Thunderbirds application process is intimidating. It's long and uncomfortable, and there was a good chance I would fail, that I wouldn't get hired. After all, there are only six pilot slots for the demonstration. But I did get hired and went on to have three years that completely changed my path. The biggest thing I learned: Have the courage to start something when you're not ready and believe in yourself enough to know that you will figure it out along the way. That is literally what every successful person you admire is doing.

If you wait for the perfect scenario, until you feel 100 percent ready, you are playing small and you will bring your progress to a crawl. The absolute worst thing you can do when it comes to making positive change is nothing. It's okay to start small.

What are your top three draws? Don't edit yourself. Just write down what comes to mind because it's what's in your heart.

THE FLIPSIDE

Writing them down is the way to start small and work toward something big.

But you're going to need some help. You'll need a wingman.

NAME YOUR WINGMAN

I'm not talking about your BFF helping you find dates in a bar—unless that's your draw, of course. I mean "wingman" as in the Air Force's definition: a pilot who flies alongside the leader of a formation to provide mutual support and backup for a mission.[5] And yes, your wingman can be a woman, and you can have more than one.

In the air, a wingman flies across from the lead jet in formation, watching their "six"—as in "six o'clock," or directly behind them—a tough place to see when you're laden down with gear. When you're leading a flight, you don't have time to be twisting around and checking behind you for threats. This leaves you vulnerable to a bad guy sneaking up on your blind spot.

Your wingman has your six, and therefore your back. Remember, fighter pilots never fly a mission on their own, and their wingmen offer backup and firepower. Your wingman recognizes when you need support and helps you prioritize obstacles that you need to overcome. They know what your strengths are and how they can fill in to help you achieve the mission.

Who are your wingmen? Who are the people in your life you can rely on to help you accomplish your mission or achieve your goal? What strengths do they see in you? How can they help you achieve your draw or support your loadout?

There's simply nothing quite like a good wingman. That's

something I learned in the cockpit during air combat maneuvers, or ACM, a mission in the F-16 that truly tested our ability to work as a team and serve as a wingman. In ACM, two jets would cruise along in formation, handling threats as they popped up. Each engagement would kick off with one of the pilots picking up an enemy aircraft visually, or on their radar, or through other indications in the cockpit. The pilot who detected the threat first would immediately direct the formation to start turning toward it, being directive and describing its location instead of the threat. That might seem strange. Why wouldn't they want to describe the type of enemy jet first? Because their priority wasn't to name the threat, but to take action to keep the flight safe by getting the formation set up in a position to control the fight.

As the threat got closer, the aircraft in the best position to engage would become the "engaged fighter" and the other would become the "supporting fighter." That way, the engaged fighter could fully focus on fighting the enemy aircraft while the supporting fighter maintained visual.

If an ACM fight was going well, it was always clear who was in what role and what their responsibilities were. Each fighter pilot knew when the other one had it handled and just needed them to stay out of the way and when they needed more proactive support.

You might think that because you'll never be in an ACM engagement these rules don't apply to you, but we can all adopt the principles of a successful wingman relationship:

- Understand each other's strengths and weaknesses.
- Clearly communicate expectations, feedback, and roles.
- Watch for threats and speak up when you see one.

THE FLIPSIDE

Think about the people in your life who see your strengths even when you don't. Who can you always rely on? Who will give you honest feedback, even if it's not always what you want to hear? Find the people who will be direct with you and exhibit the courage to have tough conversations. These are the ones who know your quirks and downfalls but can redirect you in a way that optimizes your strengths. These are also the people who can give you perspective on your inner critic when it starts to make you question your worthiness. These are your wingmen. They know your blind spots and they know how you can overcome them.

FORM YOUR BATTLE PLAN

Once you determine your loadout, list your draws, and name your wingmen, it's time to form your battle plan. When you begin with the end in mind, it will help you make good decisions along the way.

Before any mission, there's mission planning, a sort of big-picture plan that's based on the "commander's intent." That's the ultimate goal, because without that, it's hard to keep focus on the mission. You end up going down rabbit holes, further and further from the goal until you wonder, *Wait... how did we get here?* Mission planning has several components:

Commander's Intent: While most of you don't have a commander setting intent for you, you still need an ultimate goal, a true north, that you can keep moving toward. You may feel like your draws already capture this. But take this as an opportunity to ask yourself if your goal is big enough. If your inner critic is still leading the charge, you might be playing small. Next, we're going to break down your overall mission into smaller goals and objectives that feel digestible, so don't play it safe.

Know the Battlespace: Another part of mission planning is knowing the battlespace and laying out a timeline. This is an opportunity for you to put pen to paper and get some dates and times to hold yourself accountable, and also to look at obstacles that might pop up. Your inner critic has likely already brought these to your attention, but it's important to intentionally look at what a worst-case scenario looks like.

While these worst-cases usually aren't reality and can keep us stuck, I'm going to purposely tell you to think about them now. If we take the time to explore how we would handle the worst-case, much of the dread and unknown around it disappear.

The GICL: Next, I want you to consider the GICL, pronounced like "tickle" with a *G*. It's fun to say, but it stands for the Good Idea Cutoff Line. It's the point where planning ends and action begins, no matter how good the idea. There's a certain point in any battle planning where the only thing that really changes your situation is taking action, and you don't need more advice or additional ideas. Set a GICL for yourself. When that is depends on the complexity of the situation. At that point, you will ultimately make a decision to move forward with your battle plan, whatever that looks like.

The Briefing: In a flight briefing, we gather all the pilots, name the mission, and explain its goal from the start. It's the big-picture plan, followed by objectives, which are specific, measurable, and attainable goals. It's defining what success will look like. These objectives support that commander's intent we set earlier.

Specificity is key here. It should be very clear whether or not you achieved each objective.

Contracts and Contingencies: Contracts are rules we lay out before we fly the mission to keep us safe and to avoid confusion midflight. An example is, "Pilots flying east will go at odd altitudes and west at even altitudes." That keeps us from flying into each other. Contracts help you make promises to yourself before you're in the heat of the moment where your inner critic is the loudest.

If contracts are the promises to yourself, contingencies are the backup plans for things that go wrong. It's your Plan B for snafus and worst-case scenarios. You should have already explored what these might be during mission planning. Now is the time to think about how you will use your loadout and your wingmen to deal with them. It's not something to dwell on because it will keep you stuck. It's simply the answer to "What are we going to do?" if something goes wrong.

FIGHT'S ON!

After you've created your battle plan, it's time to say, "Fight's on!" It's the fighter pilot's version of "Let's go," and it comes after much preparation and training. While it's true that planning, reflection, and intention are important, *action* is what changes your life, and that's what "Fight's on" is all about. But how do you do it?

Choose three small tasks you can take toward pursuing the draw you selected and write them down. Now, pick which of the three tasks is easiest to check off. That's right: Choose the easy one first. It's a good way to ease into action. Go ahead and assign a GICL to your choice so that you don't spend too much time thinking when you should be acting.

Next, determine how your strengths can help you take action on that first task. It's probably why you chose it first. For instance, let's say you know you're good at research, so your first task toward starting your own business is to look up the paperwork and fees you'll need to make it official. If your draw is to become a stand-up comedian, your small task might be watching as many stand-up specials as you can find and taking notes, and who's not good at watching Netflix?

Bring your wingmen in, even at this early stage. They can help you get tasks done. Maybe you have an entrepreneur friend who can help you navigate the start-up process for your own business, or a funny friend who will help you decipher what works and doesn't work in stand-up. Even if your wingman has never done your draw, they can support you by cheerleading or being the wise voice of reason. But make sure it isn't someone who steps on your dreams, because that's a sure way to kill them while simultaneously feeding your inner critic.

BE BOLD. BE BRAVE.

Not everything brave I've done happened in a jet. One of them happened onstage, in three languages, only one of which I spoke.

Just eighteen months into my speaking career, I was invited to keynote a conference in Brazil, presenting to twelve hundred people. This would make me an international speaker. I imagined what it would be like to succeed at it, and I pictured the audience's excitement about seeing an American fighter pilot onstage, the expanded reach of my message, and the credibility of a big international event.

This was a major draw and, of course, I was scared. I'd never done an event in a country where English wasn't the primary language. How would I work with multiple translators, both in Spanish and Portuguese? Would the slang and the humor of my speech translate?

Naturally, I ignored my inner critic, said yes to the gig, and got to work. I asked other speakers with more experience to share advice on working with translators and foreign audiences. My friend Tim got on the phone with me and gave me specific, actionable things to make this amazing experience smooth for my client and for me.

I put in the work, cutting down my script so I could speak more slowly, allowing time for translation and reaction. I met with the translators in advance to make sure my words translated well. I rehearsed and prepared—and you know what? The audience loved it. They laughed at my jokes, and some took the time to tell me how much the speech's message meant to them. The translators even thanked me for the extra effort I'd put in to work with them.

I'd become so adept at battling my inner critic that I could see it coming a mile away and it simply couldn't derail me. I recognized and validated the draw, assessed my loadout, consulted my wingmen, and formed my battle plan. Then I got to work, and in the end, everyone was satisfied, and I became that international speaker I'd dreamed about.

Much like the military, I couldn't help but create an acronym for a list of steps, this time for keeping your critic at bay if you're B.O.L.D.:

Bring light to your critic.
Own your thoughts.
Lean on your wingmen.
Discover your power.

The first step is to recognize when your critic is telling you lies and shine a light on it, just like that stuffed animal with the creepy eyes in the bedroom corner. Next, avoid letting it take over your thoughts. Instead, own them. Start by avoiding the knee-jerk reaction of saying no to an opportunity whenever your inner critic puts the thought in your head. Instead, ask yourself these questions:

"Do I feel drawn to go after this thing?"
"What would it look like if this were to be a wild success?"
"Are my fears around this rational?"
"Six months from now, will I regret not having done this?"

If you need some encouragement or a voice of reason along the way, lean on your wingmen and remember:
You are bold! You are brave! You are not alone!
You've done the work, now it's okay to figure it out as you go. Remember, it's normal not to feel ready for big challenging things, even if you've done all the preparation. You've planned to handle what you can, allowing you a certain amount of control, but now you need to believe in yourself enough to take action. When you do, you'll win the war with your inner critic and move to the next step, action. It's the first battle in the war against an even more

THE FLIPSIDE

formidable opponent: imposter syndrome. Now there's an enemy I've worked hard to battle, and I'll share how next.

Scriptside (the old way of doing things):

- Letting that critical voice inside you call the shots.
- Ignoring your own strengths and going it alone.
- Failing to plan.
- Skipping the steps to getting started.
- Being afraid. Being very afraid.

Flipside (the new way of doing things):

- Identifying and challenging your inner critic.
- Determining your loadout, listing your draws, and naming your wingman.
- Forming your battle plan.
- Choosing three tasks to go ahead with when the "fight's on."
- Being bold. Being brave.

CHAPTER TWO
Everyone Has a Call Sign

When you're a fighter pilot in the Air Force, you get a call sign, basically a nickname, that is typically based on something dumb you did. It's not like you get to choose "Viper" or "Iceman" because it sounds cool. It's chosen for you and assigned as a rite of passage during a rowdy naming ceremony. When your call sign reminds you of a big mistake, it's humbling, and that's the point. As with all mistakes, it's only a failure if you don't learn from it, and I definitely learned a lot from mine, because it's a reference to the time I chose near death rather than failing in front of my fellow pilots.

After years of intense Air Force training, I landed an assignment in 2012 at my first operational squadron, stationed at Misawa Air Base in Japan. This was the culmination of a dream five years in the making, and my first *real* flight felt entirely surreal. I say my first *real* flight because I'd had one before it, but that had been what we call a "local area orientation," where you follow your flight lead around while they point things out. It's like a double-decker bus sightseeing tour, only in jets. "On the left is the

Appi Kogen Ski Resort. It gets great powder. And on your right, the best ramen place in town."

This flight had very low expectations when it came to actual tactics, but it did give me a chance to take in the stunning views of Northern Japan in December. Like my hometown of Medford, Wisconsin, everything was blanketed by a thick layer of snow, except the small towns dotted along the dark, angry Sea of Japan would see upward of two hundred inches of fluffy white powder per year compared to our thirty-five. As I flew alongside my flight lead, I took in how beautiful it looked, and at the same time, how foreign it felt.

My first actual flight, though, was much more challenging, and it's one I'll never forget because it follows me wherever I go in the form of the call sign I earned that day: Mace. The mission was BFM, or basic fighter maneuvering. It's what the public knows from military movies as dogfighting training. It's loosely choreographed with canned setups designed to teach us how to fly in aerial combat, a difficult skill that the Air Force continues to train even though our military hasn't been in aerial combat in a long time. Still, it helps build the basic fundamentals that a good fighter pilot needs. It creates a fast cross-check, where your eyes look for specific information. It gets you accustomed to making quick decisions in an uncomfortable and high-stress environment. It also teaches...humility.

When these training fights start, the goal is to get your air speed up and fly to the "enemy" jet's turn circle, the arc they make in the sky as they try to avoid your fire. When you're on defense, you want to make a tight turn to lose the jet behind you. On offense, you want to get as close to directly behind your target as possible so they can't lose you. So, you roll sideways—one wing up and one down—and pull back on the stick until you can point at the other jet's six, straight behind him, and "shoot" at him.

Of course, we never shoot live bullets in these drills. The guns were unarmed and "safed up" so that they were blocked from firing. But I was supposed to get in position and pull the trigger while the heads-up display would show little dots simulating where the bullets would have gone, like in a video game. If it had been real, I'd have heard and felt the "brrrrrrt" of the bullets firing at six thousand rounds per minute next to my head.

But first, I had to get behind his jet, so I rolled my airplane ninety degrees to turn, and then I pulled back on the stick at just the right time so I could take the same circle in the sky he'd just taken. My love for roller coasters had translated into a love for rolling and maneuvering the jet, so these aggressive turns didn't scare me. But then there were the "Gs," the G-forces that can get pretty strong during these turns. We all experience Gs. When you're sitting on the couch, you're at one G, and on the average roller coaster you might hit a max of four Gs, but only for a few seconds, and of course, you're just a passenger.

When you're flying a fighter jet at the turn circle entry, however, it's pretty common to hit nine Gs—nine times your body weight, nine times the force of gravity. Nine Gs hurts. The only time I've felt anything on land that's remotely like Gs was in a heavy back squat while lifting weights, when you're crouching so low you're practically sitting on the floor, a bar full of weights on your shoulders. But in the cockpit, that force isn't just on your shoulders. You feel it over every part of your body as the blood is pushed from your head toward your feet and your heart tries to pump it back upward. Once you get above five or six Gs, you can't overcome that force alone because your heart isn't strong enough. If held for too long, this can have catastrophic results.

Fighter pilots do some very specific things to combat the G-forces we put our bodies under. The first is to wear a special G-suit that looks like a pair of snow pants attached to an air hose, which plugs into the aircraft. When the jet is under the pressure of more than four Gs, the suit has internal air bladders that inflate and squeeze your lower body like a giant blood pressure cuff. I often imagined the G-suit squeezing me as if I were a tube of toothpaste.

We also do a G-strain, which is not to be confused with a G-string (an interesting miscommunication I had as a Thunderbird when I appeared on *The Kelly Clarkson Show*). The G-strain involves holding pressure in your chest cavity by taking tiny, short breaths every few seconds while also flexing your lower body muscles against the G-suit. The combination of these two actions can take a human's tolerance of five Gs up to nine or more. When the G-suit starts inflating, it's a cue to the pilot that it's time to start G-straining. The first few times you do it, you're young and inexperienced, so you have to be very intentional about it, but it soon becomes something you can do instinctively.

Both the G-suit and the G-strain are critical to survive a dogfight. Once you hit nine Gs in the turn circle entry, the aircraft typically slows down and the force drops to around seven Gs, where it's still uncomfortable, but manageable. That's because you usually lose a lot of energy as you dig your jet's control surfaces into the air to make it turn. But on this flight, my own stubbornness led to a nearly disastrous outcome.

That day, I was flying one of the Air Force's F-16s. Designed to perform at high Gs in a dogfight, it's officially named the "Fighting

Falcon," but pilots gave it the nickname the "Viper" after the fighter aircraft in the '70s TV show *Battlestar Galactica*. But I'd learned to fly F-16s in the southwestern summer heat at Luke Air Force Base in Phoenix, Arizona, and the jets don't perform as well in really hot, and therefore less dense, air. Plus, not all F-16 engines are the same. Some of them have a bit more power than others, and I learned to fly in jets with less thrust. Let's be clear: They all have a whole lot of thrust, but the differences are noticeable when you're flying near the operating limits of the aircraft.

Of course, the air in Japan in the winter is much colder than Arizona in the summer. Plus, these jets had one of the most powerful engines in the Air Force's F-16 inventory.

Still early in my Air Force career, I was young and focused on the basics. At that point in my proficiency, I was very binary with my reactions. If I saw *X*, I did *Y*. I had been overwhelmed by the complex tactics I needed to learn to fly the F-16, so I clung to the black-and-white list of steps to take as a road map toward success. I soon found out that decisions in the cockpit, rarely simple, regularly fall into a gray area.

When we started our dogfight with the radio call "Fight's on," I didn't think about adjusting how I flew to factor in the temperature or the engine. I should have pushed my throttle all the way up and lit the afterburner. The jet would have accelerated for a few seconds while I drove to the spot in the sky where I thought I'd be on the same circle as my target. I'd roll my jet, pull back on the stick, and hit nine Gs for a few seconds. I'd be able to quickly point at him, pull the trigger, get the "kill," and celebrate.

Instead, I did exactly what I thought I should do on paper. I lit the afterburner, and I felt the jet start to accelerate. But so much was happening so fast. I was flying at five hundred miles an hour, and there were so many different things to gauge: where his jet was relative to mine, the angle I was seeing as he was turning, my speed, my altitude, my Gs, and my G-strain.

As a new fighter pilot, there were a lot of things that weren't yet second nature. It was easy to become fixated on one or two of those things and lose track of the others. I was focused on just trying to keep all the balls—and my jet—in the air. I lit the afterburner and got to the turn circle, rolled ninety degrees to the right, and pulled back on the stick. I was feeling the full nine Gs, and I started G-straining.

At that point, we were both in a right turn, and as I looked up through the top of my canopy, I could see that I wasn't anywhere close to pointing at him or being near his six. He was turning tightly to outmaneuver me, and I needed to do the same. I was pulling back on the stick as hard as I could. The F-16 stick is unique. It's on the side, instead of in front of you as in other jets, and it uses a more sensitive "fly by wire" control system with electrical, not hydraulic, outputs. It works based on pressure, and pulling with twenty pounds of force maxes out what the jet can give you.

So I was trying to get my jet to carve through the air and point at him, only I was starting to feel the effects of the Gs on my body. First, my color vision started to get kind of grayish, like when you stand up too fast and end up lightheaded. But then, darkness started to close in on my sight as the blood drained from my head. This was where it got dangerous.

I could either ease off the stick or pull back the throttle and modulate power, so the energy of the aircraft would slow a little bit and the G-forces would drop enough for me to recover. But I was young and inexperienced, and I wanted so badly to prove myself. All I could think was, *You have to do this.* That was all I cared about, and so I let it go to the point where I couldn't see anything, just blackness—while I was flying through the air at hundreds of miles per hour.

I was on the verge of a G-LOC: gravity-induced loss of consciousness. And G-LOCing has killed other pilots who haven't regained consciousness in time to avoid catastrophe. In fact, it's the greatest hazard to F-16 pilots, especially because the U.S. Air Force's more recent deployed operations, lacking a credible airborne threat, tended to be relatively safe. Currently, fatalities in the jet are more common in a training scenario than in combat situations.

The Thunderbirds, which I'd join several years later, switched from the T-38 jet to the F-16 after the tragic Diamond crash in 1982, where four pilots lost their lives during one flight. Adjustments to procedures and culture led the Thunderbirds to fly with no fatalities for the next thirty-five years, until 2018, when Major Stephen "Cajun" Del Bagno G-LOCed during a practice over the Nevada Test and Training Range near Nellis Air Force Base. He had hit negative Gs in an inverted position, where the pilot pushes instead of pulls on the stick, causing the blood to rush to his head, quickly followed by positive 8.5 Gs in a pull. This rapid transition between two opposing forces on the body put Cajun in an unrecoverable position. He lost consciousness and hit the ground.

I had been stationed in Japan with Cajun and knew him as kind, funny, and extremely passionate about aviation. His accident would send shock waves through the fighter pilot community, especially the Thunderbirds. Though the Air Force has since installed a special auto-recovery system in F-16s that can detect when the pilot is pointed toward the ground too fast and too low, and right the aircraft, Thunderbird jets don't have it. We fly too

THE FLIPSIDE

close together and push the limits too close to the ground for such a system.

Six years before we lost Cajun, I teetered on the edge of my own G-LOC over Japan, and my instructor pilot was looking over his shoulder at me, expecting to see me pointing at him. Instead, he saw I was still far from being in that position and realized something wasn't right. So he said the words that stop any training scenario in its tracks: "Knock it off."

Though I could see only blackness, I could hear, and those three words pulled me back from the knife's edge of G-LOC. I finally stopped pulling on the stick, having been too stupid to quit on my own. As my vision returned, I felt a sinking feeling in my stomach. You'd think it was because, for however brief a moment, I'd put death on the table, but mortality wasn't my biggest priority. It was about doing well, so when my instructor pilot had to prematurely end the fight because of my mistake, I didn't think, *Oh my gosh! I almost died*. Instead, I thought, *I didn't do well in that fight*. My concerns were completely out of order.

In the debrief, my instructor pilot sat across the table from me and issued a stern warning: "If you do that again, you're going to kill yourself." I put on a brave face and promised that I understood and that I'd never do that again. Only then did it hit me how serious of a business I'd gotten myself into. Suddenly, I felt a wave of nausea wash over me, my throat thickened, and I felt overheated as my instructor's blunt words sank in.

Often, even when I didn't understand something, I would say I did. I had adopted the identity of being the smart girl who learned quickly and didn't ask "dumb questions." Even though I regularly struggled with various concepts as I learned the tactics I needed to fly an F-16, I rarely shared my lack of understanding, like misjudging how the conditions would shape my flight that day. More than anything, I didn't want to look stupid, and yet I'd done one of the most stupid things a fighter pilot can do.

The only person who had actually witnessed my massive misstep was my instructor, but I knew how it worked. A story like this would quickly circulate among the other pilots, and by the time I was back at work the next day, I figured it would be common knowledge that I wasn't good enough for this career. Everybody in a squadron knew when pilots were crushing it and when they were struggling, and if word spread that I couldn't cut it as a fighter pilot... well, that felt to me like the worst thing that could happen.

That night, I went home for dinner and stared into the pot of water on my stove as I waited for it to boil. My thoughts were consumed by how poorly I had performed that day. As I mentally spiraled, I leaned against the kitchen cabinets, slid to the floor, and began to cry. I wondered if I had what it took to be a fighter pilot. I had so much invested in being there, and I owed the Air Force a ten-year contract because they'd spent so much money on training me. If I couldn't be a fighter pilot, the Air Force would decide where I'd end up, and chances were whatever I wound up doing wouldn't have been as extraordinary as flying F-16s. And more important than that, I'd be letting down everyone who'd expected me to excel at this. I was only about a year and a half in, and I already had an epic fail on my record. I felt trapped and sure that my reputation in my new community was already in the gutter.

In hindsight, I was probably average for a new fighter pilot, and there were plenty of other pilots who made similar mistakes, but I had no perspective on it. Still, I thought that quitting would make me an even bigger failure, so I stuck it out even though I knew I wouldn't be able to put that mistake behind me. The Air Force, and specifically the fighter pilot culture, would make sure of it.

THE BLESSINGS OF MISTAKES

Only fighter pilots who've already received their call signs are allowed to be a part of the call sign–naming ceremony, a rite of passage that happens a few months into a new pilot's time at their first combat unit. So, my forty or so colleagues kicked me and the handful of other new pilots out of the room. By this rite of passage, fledgling fighter pilots are officially certified as combat-qualified wingmen. You've completed dozens of flights during Mission Qualification Training—and dozens of flights for a new pilot means there are plenty of mistakes to use as fodder to name you something embarrassing.

There's a big whiteboard involved and plenty of beer, and your teammates share stories of your mistakes, throwing out potential names related to them. The stories range from 100 percent true to greatly exaggerated, theatrical retellings for the entertainment of those in attendance. We based our storytelling on the 10 percent rule: At least 10 percent of the story must be factual. With 90 percent leeway, these tales can get out of hand.

I was later told that they didn't have that many stories about me, but that flight where I nearly G-LOCed stuck out. So, they named me MACE, an acronym for Mach at Circle Entry. In other words, I had flown my jet so fast during a maneuver that I went supersonic, the speed of sound. I was flying much faster than I should have been, and the jet had so much energy that when I rolled and pulled, I hit nine Gs and then held it for a full 360-degree turn.

Sounds cool, right? I mean, unlike some other pilots, I was lucky enough to avoid having my call sign reference having a bathroom emergency midflight, but I have been in formation with others who faced these dire circumstances. The report of their "mission's success" made me laugh harder than I'd ever laughed in the cockpit:

> "I had to sacrifice my flight gloves, a map, and my helmet bag, but we got it done."

But my call sign, Mace, which I'm not going to write in all caps here because it makes me feel like I'm yelling, isn't based on something embarrassing so much as borderline dangerous. It's what happened *after* I went supersonic that's important: I didn't immediately back down from a potentially fatal mistake. My fear of public failure led me to this poor choice, which now follows me wherever I go, because I'm not just Michelle Curran, I'm Mace Curran.

In the cockpit, I thought I had to be like Maverick, and that could have been catastrophic. So I learned how to be Mace instead, and Mace had a lot to learn from her mistakes.

LEARN FROM YOUR CALL SIGN

So, what's your version of nearly G-LOCing? It doesn't have to have such serious consequences, but it should be something you need to learn from. It can be a single mistake (maybe one you'd rather not think about), or a series of mistakes that you still haven't learned from—bad jobs, bad relationships, or bad choices.

If we locked the people you consider your wingmen in a room

with a whiteboard and some beer, what would they end up naming you based on these mistakes? And what would get you to knock it off?

I wasn't ashamed of my call sign, but for a long time, I was ashamed of nearly G-LOCing. Sometimes I'd tell my call sign story up to the part where I broke the speed of sound and leave out the detail about nearly passing out, because it made me sound tough instead of reckless. But now I wear "Mace" as a badge of honor because I can use it to teach other people to knock it off. Over the course of my flying career, I came to see that mistakes actually benefit us, but only if we take time to learn from them. In school, we're not taught to do that. There's no debrief in fifth-grade math class.

Columbia University's Center for Teaching and Learning reported that corrective feedback, including an analysis of the mistake, helps students learn better, but that's not how our educational system is set up.[1] The researchers said that "error avoidance" is the rule in most American classrooms.[2] How can you learn from your mistakes in life if you're taught to avoid them in school?

But the U.S. military understands the value hidden in mistakes. That's why we debrief, reviewing our flights immediately afterward compared against the briefs we'd done preflight. You can't effectively debrief if you don't brief first. The goal I'd set for BFM that day was to pull behind my instructor pilot's airplane and fire my (unarmed) gun at him, and not only hadn't I achieved that, I'd flown so fast and put my body under so many Gs, I couldn't even see him—or anything. Fighter squadrons take it to an extreme level, and it's generally accepted that much more learning takes place in the debrief than during the actual flight, as I'd soon find out.

Here's how debriefing works: Everyone who flew in the mission goes into a room and shuts the door. No phones, no interruptions, no distractions. We then list the failures from a flight on a whiteboard, and dig into all the details, such as location, time, and issue. We review the contributing factors, starting with the plan and the preparation, before going into the details of any errors. We continue to peel back the layers of why something went wrong until we can get down to the true root cause. Was it a lack of information or a lack of understanding? Or did the pilot have information and understanding, but still made the wrong decision?

Everyone, no matter their rank, is held accountable for their errors, but we stick to the facts. We talk about what happened and why, not about how it made us feel. There is an understanding that we own up to our errors to make all of us better pilots, and that the only way to do that is to create a culture where lessons learned are valued above the opportunity to shame someone for their mistakes.

So when my instructor pilot told me in my debrief in Japan, "If you do that again, you're going to kill yourself," he was stating a fact. The truth hurts, but that's the point. Note the mistake, learn from it, and knock it off. I'd eventually come to see the debrief as a place to put aside shame or embarrassment and learn valuable lessons from my mistakes. And making time to do this cross-examination is just as valuable outside the cockpit.

Now, you may think, *Mace, I'm not going to brief and debrief a first date.* Well, why not? If you write down what you're hoping to get from the date, it will help you determine whether it was a success. For example, you may want to avoid going home with them on your first date or you may plan on splitting the check, even if they protest. These are quantifiable, debriefable actions to take. Without them, you may let subjective factors muddy your assessment of how the date really went—and the next thing you know, you're putting away your credit card and climbing into an Uber to your date's place just like you said you wouldn't.

Let's say you've made a series of mistakes while swiping right on dating apps, ignoring the red flags and focusing only on physical attractiveness in their profile picture. If you go into a first date doing the same thing but expecting different results, you'll likely be disappointed. So why not set goals and then determine whether you've met them? Why not take a hard look at what went wrong and what went right? You don't have to bust out your whiteboard, but intentional reflection can pay dividends. Instead of hiding from your mistakes, use them to course-correct your life.

THE SCIENCE OF MISTAKES

Our brains alert us pretty quickly when we've made a mistake with that sinking "uh-oh" feeling that comes over you like a wave. Scientists call the alert an "early warning signal," and its purpose is to prevent us from repeating our mistakes.[3]

Yet making a mistake can actually benefit us if we're given a chance to learn from it. Researchers at Michigan State University found that people who see mistakes as a wake-up call and do the work of determining what went wrong come to view them as lessons. After we make a mistake, the brain creates two quick signals within a quarter of a second. The first is what the lead researcher calls the "'oh crap' response," like when I realized that my vision was starting to fail during the G-strain in the middle of the maneuver. The second shows whether the person is aware of the mistake and is trying to fix it. The study participants who believed they could learn from their mistakes bounced back better and improved their performance on repeat tests.[4]

It took me a while to accept that.

I thought that the solution to pushing through such a big mistake was grit, which researcher and author Angela Duckworth defines as sticking with something until you master it. Her research on students found that when it comes to success, grit matters as much as intelligence.[5] She discovered that smarter students actually had less grit, and that less intelligent students compensated with determination. Guess what? The students with grit ended up with higher GPAs.[6]

The military puts plenty of stock in the concept of grit. At the United States Military Academy at West Point, a cadet's grit score helps predict whether they'll survive the tough training program leading into the first year called "Beast Barracks." It's that "never say die" attitude that builds perseverance.[7]

But there can be a price to pay for pushing forward. Though we've all been taught that "winners never quit," you have to know "when to quit and when to grit," as Harvard psychologist and *Emotional Agility* author Susan David said. "We should be gritty, yes, but not stupid."[8]

When I'm pushing people for boldness and grit, I'm not saying you should blindly forge ahead with no risk assessment. In the cockpit, we are constantly taking *calculated* risks—not the kind of risks that increase the odds of getting us killed. My near G-LOC is a good example of doing it all wrong. I made a huge misassessment of whether the risk was worth it. Of course, it wasn't worth risking my life just to avoid embarrassment in front of my team.

The thing is, when you're making a mistake, the brain thinks it's making the right decision. Otherwise, why else would you do it? Science has an explanation for this. A study at Carnegie Mellon University found that the brain makes mistakes because it applies incorrect inner beliefs to the action you're taking.[9] In other words, your inner thoughts and beliefs don't match reality. So while I was G-straining, something inside me was saying, *I've got to get him, no matter what*, even though "no matter what" proved to be exceptionally dangerous.

It's a bit like summit fever, which is part of the reason why there are some two hundred dead bodies near the summit of Mt. Everest. Summit fever happens when you're so invested in the outcome of a goal, you make bad decisions. It's easier to take dangerous risks when you've put that kind of commitment, money, and effort into a goal.

At any point during that fateful flight, I could've decided to adjust my speed and still remain in the dogfight. When I realized that pulling relentlessly on the stick was no longer serving me and my real objective, I could have changed my tactic. My fate wasn't sealed during my initial mistake of going too fast before I rolled and pulled. All it would have taken was a minor correction. But my lack of experience and my misguided grit led me down a path that could have ended in catastrophe.

During my near G-LOC, I was under all the physical demands that come with a high-G maneuver, so the pressure to perform was really just one factor. But in that instance, that factor might have been enough to lead me to push things as far as I did. I might not have known what I didn't know so early in my career as a pilot, but the most important thing was that I learned to knock it off.

If you maintain some perspective during a mistake and are willing to recognize your earlier errors, you can catch yourself before you're about to G-LOC. If your actions are preventing you from achieving your goals, stop pulling on that stick!

SHAME VS. GUILT

For a while, I was ashamed of my big mistake, and that's probably what got in my way of immediately using it as a valuable lesson. I felt shame, not guilt.

In a *PLOS One* study, researchers found that shame typically leads people to hide from their mistakes, while guilt tends to drive people to amend them.[10] Though some people use the terms "shame" and "guilt" interchangeably, they're different feelings. People who feel guilty about a mistake can separate themselves out from the action, but people who feel ashamed see the error and themselves as one and the same.

Brené Brown has based an entire career on shame research, and her TED talk on shame has about seven million views. She calls shame an "unspoken epidemic,"[11] and it's not the same as guilt. To her, guilt is "I did something bad," but shame? That's "I *am* bad." She says that for men, shame is usually tied to being perceived as weak, while for women it's about conflicting expectations of who we're supposed to be.[12]

I'm no psychologist, but as a fighter pilot, I was a woman doing what has historically been a man's job—currently only 5 percent of fighter pilots are female. When I first started out, it was half that. I was trying to prove myself to the other pilots, who were all men, in a job that few women had done before me. I had a double whammy of shame about my perceived weakness and about conflicting expectations for me as a woman.

THE FLIPSIDE

When you feel shame, the brain reacts like it is facing physical danger with the flight-fight-freeze response. Your face may flush, and you may feel disconnected from your body or the situation. That's because shame is actually a protective response, reminding you that you're not safe. If, for example, our parents yelled at us for nearly darting into traffic when we were little, the feeling of shame was designed to protect us from doing that again.

I'm sure I felt the effects of shame in the debrief of my near G-LOC. It's probably the reason why it took me so long to work my way through the feelings associated with my mistake. But within a few months of receiving my call sign, I embraced it. I didn't really have much of a choice—it was the only thing I was called by every single person I interacted with in Japan.

To get there, the first step was recognizing that everyone makes mistakes and that mine was actually not an uncommon one for a new fighter pilot. Then I began to treat younger me, the inexperienced fighter pilot, with kindness. Next, I reframed shame as guilt, telling myself that the incident didn't define me. Finally, I worked through the guilt by forgiving myself. It sounds simple, but it can be hard to do.

That day during BFM, there was so much for me to keep track of at once, but I was new at much of it. I was in training to learn the basics of dogfighting; it was a different kind of jet in different conditions than I was used to, and I was in an entirely unnatural maneuver that put so much pressure on my body, I could barely stay conscious. My blood had flushed out of my brain and toward my feet because I was flying at the speed of sound! When I laid it all out there, I realized how extreme these circumstances were and that feeling shame for not performing perfectly now seems ridiculous.

We all make mistakes. Your mindset around those mistakes, especially the life-changing ones, is what will be the difference between getting stuck in the errors and learning from them. Going forward, you may want to change your thinking and your habits. For me, that meant no longer trying to prove myself to others, especially if it meant putting myself in danger.

I don't mean just in the cockpit. On the ground, I also had to learn to stop and evaluate why I was pouring on the grit to prove myself to my fellow fighter pilots. If the reason didn't serve my ultimate goals, I learned to course-correct back to something that did.

I've also needed to do this outside the military. Now that I'm an entrepreneur, I recognize that the job comes with a similar "grit it out" mentality, the idea that you need to prove you're strong enough to make it in the business world. This can lead to saying yes to everything and everyone, because you never know which interactions will lead to a huge opportunity. Only those willing to lay it all out there will succeed. It's the epitome of the hustle culture glamorized on social media. No days off! Boss babe! Every day I'm hus-hustlin'! Even after all I learned in my first career, I found myself being pulled toward this same mindset.

A few months after leaving active duty, my breakfast sat cold and half-eaten on my desk at 3:00 p.m. I had been on Zoom calls all day with no more than fifteen minutes between them, for hours. These were random networking calls from people who'd contacted me through LinkedIn, including someone who wanted to pitch a fighter pilot movie and someone else with an offer to consult for an AI-based, pilot-training tech company.

I felt this inner conflict: I should be grateful for all these opportunities, but I was exhausted. It took a mentor asking me what my goals really were and if the busywork I was doing served them to act as my call to knock it off. Only then did I realize that I needed to work smarter, not harder, just like in the cockpit.

This mentor told me that if the opportunity didn't fall under passion, profit, or politics, it should be a "no-go." Another told me if it's not a "hell yeah!" then it should be a no. It still took some time to get better at identifying the hell yeahs. Then I could let go of any guilt or shame I had around not pleasing everyone.

After years of self-reflection and reframing my thinking, I'd learned that a failure is its own kind of success if you know it'll lead to long-term improvements. Now I'm hoping to show you how to live a fuller life based on what I've learned in the cockpit and on the ground.

I'm proud of that work. I'm proud of the inspiration I got to create while flying for the Thunderbirds, and I'm proud of my call sign. Finally.

Scriptside:

- Letting shame block you from learning from your mistakes.
- Failing to review what went wrong.
- Not figuring out how to fix it.
- Not defining what's worth fighting for.

Flipside:

- Learning from your call sign.
- Debriefing your mistakes.
- Course-correcting yourself.
- Believing that if it's not a "hell yeah," it's a "no-go."

CHAPTER THREE
Stop Playing Dress-Up

Throughout my military career, my inner critic liked to remind me that I didn't fit in. Often, I was the only woman in my squadron, the only one with a braid hanging out from the back of my flight helmet. Believe me, this fact didn't slip past my flight commander in IFF, Introduction to Fighter Fundamentals, a course designed to teach us freshly assigned fighter pilots the basics of flying fast jets. It's a class of mock dogfighting, studying enemy threats, and exposure to fighter pilot culture.

This includes roll call, a Friday night gathering of all of the pilots in the squadron bar, a tradition that stems back to World War II when its somber purpose was to count the pilots and crew who'd returned, or not returned, from their missions. Though it's designed to foster camaraderie and embrace tradition, for decades after the war it had devolved into a raucous event with flowing kegs and sometimes, bawdy songs. Think of Tom Cruise singing, "You've Lost That Loving Feeling" with a throng of pilots as backup singers in the original *Top Gun*, and you've got just a little taste of the rowdy vibe.

Roll call was almost like an indoctrination, and we were the

THE FLIPSIDE

lucky few, just a handful of new pilots who had been selected from among our various training classes to fly fighter aircraft while most would go on to fly "heavies," aircraft used for carrying cargo and air refueling. For those of us who dreamed of flying fighters, the odds hadn't been in our favor, but here we were, the pilots that many in the classes behind us envied. We'd beaten the odds. It gave me a temporary feeling of success and belonging, and I held my head high as I walked the halls of the training squadron waiting for my turn at IFF to start. Meanwhile, I got a nice reprieve from imposter syndrome that had occasionally begun to rear its ugly head.

Back then, I was young and naïve, looking up to our instructors, veteran fighter pilots who'd seen combat in places like Afghanistan or Iraq. To us, they were gods. So imagine how I felt when my flight commander and immediate supervisor pulled me into his office to deliver a warning right before my first IFF roll call. He was an intimidating major two ranks higher and a decade older, a seasoned fighter pilot who was hanging on to the "boys will be boys" culture of roll call tradition, including singing raunchy fighter pilot songs that would make you blush and participating in other loud and drunken debauchery. Back in the "good old days," strippers were sometimes brought in, and the pilots occasionally had fistfights with one another. At that time, there were no women in the squadrons, and now, here I was, the only one in ours.

"We're going to have roll call on Friday night, and there's going to be dirty jokes and some things you might not be comfortable with," my commander warned me. "And if I wanna put girls with bouncing boobies on a trampoline on the TV, I'm going to do that. If you don't like it, leave."

He really did say the words "bouncing boobies." A grown man. In the twenty-first century, from his desk chair in his office. *At work.*

Though women had been allowed to become fighter pilots for nearly two decades at that point, the culture was still... "evolving." But this was our *workplace*, a professional environment. I'm all for work hard, play hard, but everyone would survive without the "bouncing boobies."

It was one of the first big indications I got that I didn't naturally belong in this community, and it wasn't the last. During our F-16 training, we were assigned silly temporary call signs for the duration of the course. Naturally, there was a naming ceremony, and for our squadron, the Emerald Knights, the commander "knighted" us with a sword. Each pilot would get called forward and then we'd take a knee so the commander could touch each shoulder with the sword and announce our temporary call signs. When it was my turn, I took a knee. Suddenly, I'm pelvis high, and it didn't go unnoticed.

"Whoa, whoa," the commander laughed. "What are you doing down there?" Laughter erupted from the mostly male class and instructors. He knighted me "Candy," because I loved snacking on gummy bears while studying in the flight room, and someone thought it sounded like a stripper name. I stood up, red-faced.

After that night, I became a chameleon. I played dress-up because being myself wasn't working. When I showed up as "one of the bros," I even overheard one of my pilot training classmates tell another student, "It's been great. Luckily, she's super chill... Yeah, I know, it could have been a pain in the ass."

Again and again, I'd receive reinforcement that the only way a female pilot like me could excel in the USAF was to avoid making waves or making anyone else feel uncomfortable. For a while, "chill" was valued, but as I moved into high-performing environments like IFF and my first combat squadron, it was the loud, overconfident, and at times crude pilots who were rewarded. So I had to learn to adapt.

I can't tell you how many times I got the feedback, "You need to be more assertive, more confident. If you want to be a fighter pilot, you need to act more like one." The men would be rewarded for confidence even if they were incompetent, and if I didn't act the same way, I was told I didn't belong. Yet we all know what happens when women appear overly confident, and it's rarely followed by accolades.

I had to figure out how to fit in and be one of the guys. Though I didn't like sports, I pretended I did because it was regularly a topic of conversation among my classmates. They'd talk about football and I'd add simply "Yeah, Green Bay Packers" because I was born in Wisconsin, but really, I couldn't have cared less. I just wanted to be included.

Worse, I'd laugh at jokes that made me uncomfortable, and I wouldn't speak up if my colleagues did or said something that bothered me. I would observe how everyone else was reacting to a joke and attempt to smile or laugh enough that I didn't stand out. Inside though, that uneasy feeling in the pit of my stomach creeped in. I decided that flying under the radar would be my priority, and that sometimes meant letting cultural expectations supersede my values. I had to choose between being one of the guys or being myself. To fit in, I couldn't be authentic, and if I couldn't be authentic, I couldn't be me.

It was a recipe for imposter syndrome, a phrase coined in the 1970s by two female psychologists who'd sought to explain why high-achieving women sometimes felt like frauds.[1] They surmised that "despite outstanding academic and professional accomplishments, women who experience the imposter phenomenon persist in believing that they are really not bright and have fooled anyone who thinks otherwise."[2]

I had felt this way occasionally during earlier training. At times, after a flight went particularly well, I thought, *Okay, you fooled them again today.* I wanted to be part of the group, to belong. But how could I feel like I belonged in a culture that ranged from not knowing what to do with me to not wanting me to be there at all?

This atmosphere created an environment where no one, including the men, felt like they could show weakness. When I asked a fighter pilot who'd already spent a year stationed in Misawa, "At what point did you feel like you knew what you were doing?" he replied, "I'll let you know when I find out."

I was so grateful for that vulnerable sliver of camaraderie that I began to tell that story at his going-away party—but I didn't get far.

"Oh no! Please don't share that," he interrupted. He'd spent three years flying in that squadron, and he was mortified that I was about to reveal that he might have felt like an imposter at some point.

POOPY SUITS AND PIDDLE PACKS

My colleague with the call sign "Smokin'" and I were the only two women in the squadron of would-be F-16 pilots when we were pulled aside before our first roll call of the training. This time, it wasn't a "bouncing boobies" moment, but it had been merely six months since that particular awkwardness. This time, our commander still had a warning about roll call's rowdiness to deliver, but his was more of a heads-up, a friendlier "Hey, this is gonna happen so be ready for it."

THE FLIPSIDE

"It" turned out to be fighter pilot songs focused on the theme of demeaning women, constant sexual innuendos, and, occasionally, over-the-top remarks that even made many of the men noticeably uncomfortable. He never said, "We're going to stop doing this now that you two are here." Or, "We're working to change this, so please be patient with us." It was up to Smokin' and me to adjust because many of our colleagues had resisted the fact that combat cockpits had (long) opened to women, and change was slow.

Even the fighter pilot culture's language was gendered. In classes, instructors would say, "Okay, gentlemen..." Sometimes, they'd catch themselves and add, "and *Candy*." Frankly, that was worse because I didn't love being singled out. They'd even do it for the good stuff, like when I was celebrated as extra impressive when I did well, a kind of pat on the back for doing a good job "for a girl" as though they had expected less from me. Even when I did well, I couldn't win. I didn't know how I should react. Part of me was proud of my successes and the other part worried about what my peers thought. Once, after I'd crushed it in a high-stakes flight simulator test, not only earning praise from my squadron but winning an award for bravery under (virtual) fire, I found myself checking the other brand-new wingmen's reactions. I couldn't just celebrate the moment. I had to worry that my win would make them feel less manly, ultimately affecting our relationship and my standing in the squadron. It always felt like I was walking a tightrope and never finding my footing.

It was a lot of little stuff, things they call microaggressions now, that made me feel singled out, and systemic issues. As part of our flight gear, we pilots wore anti-exposure suits, which we called "poopy suits," because they lacked rear zippers for when nature calls. Worn over our underwear and T-shirts, these waterproof suits would keep us alive longer if we ejected into the freezing waters of the Pacific. But there was no women's locker room, so Smokin' and I changed into our life-saving gear in the tiny bathroom. Lacking space, we would occasionally end up with a flight suit sleeve dipped in the toilet or getting hit by the door if anyone else tried to come in.

Then there were the piddle packs for urinating while on long flights in the cockpit. The men could just pop their piddle packs open and go, but this presented a big problem for us women because they didn't fit our anatomy. We had to find our own solution, so we ordered our own, which were more suitable and designed for camping, from an outdoor store. The Air Force occasionally stocked these modified piddle packs for us but would regularly run out. We had to ensure we had our own stash on hand so we could focus on what really mattered, like flying the jet and completing the mission.

That wasn't all. The flight suit zipper stopped mid–pubic bone, which works for guys who need to pee, but not for us. So the Air Force started making "Mrs." flight suits with a zipper that extends all the way to your butt, allowing us to do a midair squat and let gravity do the rest. But they'd constantly run out of stock because of a "lack of demand." Well, yeah. If only a handful of fighter pilots are female, the demand won't be high, but this was gear we needed to do our jobs.

While it didn't affect our usual, day-to-day training flights, long flights were another story. When we were piloting cross-country hauls across the ocean to move jets to a deployed location or running four-to-six-hour combat flights, the stakes were high and the missions complex. Worrying about when and how you will

THE FLIPSIDE

inevitably use the bathroom added a layer of stress that elevated the risk even higher.

The Air Force had spent millions of dollars on our training for what we were repeatedly reminded was an elite, highly valued career, and yet they failed to support us in such a basic, biological way. It said to me that the system wasn't built for us and, worse, the system wasn't willing to adapt for us.

It still isn't. Even now, we veterans pass down our Mrs. flight suits to new female fighter pilots. I sent mine to an F-15 pilot who happened to be my size, because I didn't need them to fly for the Thunderbirds. A solid quarter century after women first became fighter pilots, it was easier for her to get my secondhand flight suits than to secure new ones from the Air Force.

But the worst part of being singled out was the blatant, though rare, hostility that stuck with me for years. The first time any of my classmates highlighted gender in a negative way caught me off guard. Up until that point I hadn't thought much about it, but that would rapidly change as it became impossible to ignore.

It was at the very end of pilot training, right before we were taking our last check ride and a few weeks before we'd find out which aircraft we'd receive. Six of my classmates, all men, were also in the running for the handful of fighter aircraft our class would likely get, and I'd put the coveted F-16 at the top of my list. That was when one classmate and I realized we were neck and neck for one of the top spots in the class, fighting for the fighter. As we sat together preparing for this final check ride, he said, "I don't know why I'm even trying, because they're going to choose you for a fighter to 'check the box.'" In other words, I would be a diversity hire, so I'd get my first-choice fighter jet, not because I'd worked my ass off, consistently rising to the top in my class, but because of my gender.

This was when I really started to feel like I was under a microscope and that everyone was watching to see if I'd fail. If I didn't, then all female fighter pilots would be considered a good thing. If I did, then they'd all be a bad thing.

No pressure.

But in time, I learned that I'd blown that out of proportion, making it way more of a factor than it was in reality. Most people were more likely worried about themselves. And the few who did care did so because of their own insecurities. When they were worried that they weren't performing well enough to be at the top of the class, I was an easy target.

After I'd become a more experienced fighter pilot, an instructor pilot, a flight commander, and the Lead Solo of the Thunderbirds, I figured out an important fact: No matter where you are in your career, you're mostly worried about your own performance. People aren't really thinking about you much at all—and if they are, it has nothing to do with you and everything to do with them.

IMPOSTER SYNDROME: NOT JUST FOR GIRLS

Plenty of high achievers have confessed to feeling imposter syndrome. Lady Gaga said in her HBO special that she still feels like the "loser kid" in high school, and she tells herself every morning that she's a superstar so she can be the best of what her fans expect of her.[3] Real estate mogul and *Shark Tank* star Barbara Corcoran confessed in an Instagram post, "Even when I sold my business for $66 million, I felt like an absolute fraud!"[4] In fact, she also said, "The more successful someone is, the more self-doubt they have, because that's what drives them." The super-successful Meryl Streep, who has the most Academy Award nominations of any actor, of all time, said that a little voice in her head has wondered, "Why would anyone want to see me again in a movie? And I don't know how to act anyway, so why am I doing this?"[5]

As organizational psychologist and bestselling author Adam Grant tweeted, "Impostor syndrome isn't a disease. It's a normal response to internalizing impossibly high standards."[6] Who does that? High-achieving people. *Harvard Business Review* explains that the imposter syndrome we know now as the affliction of #ladybosses is a "fairly universal feeling of discomfort, second-guessing, and mild anxiety in the workplace" that was pathologized, particularly for women.

Imposter syndrome isn't a women's issue, but we're made to feel like it is. *Harvard Business Review* adds that genderizing it puts the blame on individuals with no acknowledgment of the history and culture that feed it: "Imposter syndrome directs our view toward fixing women at work instead of fixing the places where women work."[7]

In fact, I am still floored when men come up to me after I deliver a keynote to confess that they, too, suffer from imposter syndrome. These are usually white, heterosexual men—often CEOs—admitting to feeling like frauds. I had no idea that accomplished men also felt that way, and I imagine that they must suffer in silence. Where's the men's magazine exposé on that? Just once, I'd like to read the headline, "How Boy Bosses Can Overcome Imposter Syndrome." This #ladypilot thinks it's time we eject it from our conversations around only women and work toward changing the culture that feeds it.

On his podcast, *The Accidental Creative*, author and speaker Todd Henry took the Three Ps of Optimism created by Martin Seligman,[8] known as the father of positive psychology, and applied it to imposter syndrome.[9] Seligman had said that the three qualities—personalization, pervasiveness, and permanence—were crucial to rebounding from hardship. Henry added that the story we tell ourselves throughout these three Ps defines our reality and, therefore, our feelings of imposter syndrome.

"So instead of, 'I failed,' the narrative says, 'I'm a failure,'" he said. That's the personalization part, which we also discussed as the difference between shame and guilt. Then, he added, we spread it everywhere in our lives: "I'm probably going to fail at everything." That's pervasiveness. Permanence means we're stuck with it for life, as in, "I'll always be a failure." He said we have to recognize how these three qualities are seeping into our lives and pull them like weeds.

Imposter syndrome keeps us stuck. When you feel like a fraud, you tend to avoid going for that promotion, swiping right on that attractive person online, applying for that grant, going for an advanced degree, or trying anything remotely out of your comfort zone. And that's too bad, because that's where the growth happens.

Imposter syndrome takes you down a path that doesn't serve you.

I'm a failure → I'll always fail → I'll forever be a failure

When it should be:

I failed → I'll learn from this mistake → The lesson I learn will help me grow

But you have to be willing to risk making a mistake in the first place.

WE'RE ALL BEGINNERS AT SOMETHING

There's a certain amount of imposter syndrome to be expected when you're starting something new and different. You're a beginner, and you don't have a proven track record of success, so you can feel like you don't belong.

Throughout pilot training, I had worked hard, embraced all I needed to learn, and excelled. But at some point along the journey from graduation through IFF and the F-16 course and into my first combat unit, the normal level of discomfort and apprehension that comes along with being new began to mutate into much more.

I began to feel like I couldn't show weakness in that environment, so I didn't ask questions. Instead, I studied my butt off until super late at night, in private. Everyone else was new, too, but I thought they knew things that I didn't, and I was overwhelmed with everything I had to learn to go from flying the F-16 to operating it at the tactical level required to be a fighter pilot.

Add in the good old boys' culture full of bravado and those raucous roll calls, plus the fact that I was very much in the minority, and I believed I had to be a certain way to be welcomed in this community. I even worked hard to lose my Wisconsin accent in pilot training, because I got made fun of all the time. I actually practiced pronouncing words like "bag," "flag," "about," and "house" without an accent, but I couldn't practice losing my gender.

I am of the mindset that giving one another a hard time builds camaraderie. It's often a sign that you've been accepted into a group, and I participated in it many times. But that requires one big underlying thing to be true: You have to feel like you belong. Without that foundation of belonging, the humor of these otherwise well-meaning jabs gets lost as they slowly chip away at your identity and confidence.

As if it wasn't enough to have pressure from the outside, my inner critic harassed me from the inside. It amplified my external struggles by attacking my personal character because the inner critic never supplies constructive criticism. It calls you an idiot and tells you that you're incompetent, and that rapidly becomes the truth for you.

But if there's one thing I've learned as a fighter pilot playing dress-up to fit in, it's that you have to shine a light on the inner critic before it fuels imposter syndrome out of control. Thoughts in your head are not always factual, and they don't always reflect reality. They're our perception of them, and our perception is often skewed.

We're not imposters. We're *beginners*. Even Michael Jordan took the beginner's mindset well into his basketball career.[10] After a three-peat of NBA championships with the Chicago Bulls, he played baseball for the White Sox's minor league team, the Birmingham Barons, in 1994. Playing professional baseball had been a childhood dream, and he was in the throes of grief, having lost his dad, so he signed with Major League Baseball to test out a new career, even playing in an exhibition game with the Chicago White Sox.

The sports press was not exactly kind, with one sportscaster calling him "a fish out of water." But Jordan remained humble—as humble as you can be when you own your own plane. Jordan told the sports press at the time, "If I ever develop the skills to be up here [on the Chicago White Sox team], great. If I don't, at least I fulfilled a dream at least trying."[11] His training partner said, "Here is the greatest basketball player of all time, and he's looking at me to say, 'Teach me.'"[12] The entire country was watching all six foot six of him, swinging his bat in a brand-new, bright white baseball uniform, and yet, he didn't succumb to imposter syndrome because he had the beginner's mind.

It's not easy to do. Expect a learning curve, and, if it makes you feel better, catch the naysayers off guard by beating them to your "inconvenient truth." It's something that I heard on a Mel Robbins podcast interview with her daughter, Kendall, who shared a story about the imposter syndrome she felt while attending a music festival as a newbie to the songwriting industry.[13]

Kendall had the great fortune of getting a special pass to a backstage tent. There she ran into some of her Grammy-winning songwriting idols—and instantly felt like a fraud. A new singer-songwriter, she didn't have a Spotify list or even an Instagram account to direct people to.

"I was so uncomfortable and so just embarrassed and just felt like, why am I here?" Yet she'd recently graduated from one of the country's top college programs for pop songwriting. She wasn't a fraud. She was just new. So she admitted to everyone she met that she was a beginner, relying on something a professor had taught her: You become one of the most powerful people in the room when you beat everyone to your "inconvenient truth," your glaring weakness. Instead of pretending you have it all figured out, tell them the truth. For Kendall, it was that she had no music out, no fans, no following, and no gigs.

Then she turned her inconvenient truth into a plus by saying she was excited to learn from the accomplished songwriters at the event. That changed everything, because suddenly she had gone from apologizing for her inexperience to becoming the apprentice that everyone wanted to help.

The calculus for imposter syndrome is simple, but insidious:

high achiever + new challenges x self-doubt = imposter syndrome

Beat them to the equals sign by embracing the fact that you're a beginner and then say your inconvenient truth out loud. That way, you disrupt the calculus by announcing that you, like Michael Jordan, are trying something new, so you might take a few missteps along the way. On this particular task, you haven't yet achieved your ten thousand hours, so you haven't reached mastery. If you could hang a sign, it might say, "Please pardon me while I make important upgrades."

I imagine it as a scene where someone is trying to blackmail me:

Criminal: "You better do what I say or I'm going to tell the whole world you aren't good at [insert skill]."

Me: "Cool. Everyone already knows that since I'm new."

Criminal: *(walks away, defeated)*

Say it so even your inner critic will listen and back down.

I've learned to interrupt my inner critic's ramblings, which are nothing more than negative thought patterns, and divert their path. For instance, early in my speaking career, I had a gig before an audience of eight hundred sheriff's department employees. It was an intimidating group of mostly guys, who appeared very confident. It was a sea of uniforms. Yes, I'm a fighter pilot, and they looked up to me for that, but at that moment, I was full of self-doubt because I was about to talk about a warm fuzzy topic—vulnerability—something that I wasn't sure would resonate with them.

But then I reminded myself that they'd hired me for a reason. What I was about to talk about wasn't a secret, and they'd all showed up for it. After a successful Air Force career, I had the credibility to be there, and they'd picked me to speak at their event. So I took the stage, and you know what? It went well. Many of them came up after to thank me, admitting that it wasn't a topic they openly talked about but experienced quite a bit.

"I felt like you were speaking directly to me," several said. I'd been scared that my message wouldn't resonate with them, and yet it not only had, they were inspired enough by my speech to tell me so.

Now I no longer hear my inner critic as much before I take the stage because I'm not a beginner anymore. But early on, I had to intentionally interrupt my thought patterns, redirect them, and remind myself why that wasn't the truth and why I deserved to be there.

So did my friend Katie Boer, who even today carries note cards around in her purse to help her endure the tough times. She said that when she was going through a challenging period as a small-market television reporter, she felt like an imposter. She was a cub reporter in a cutthroat industry, and it made her feel isolated and exhausted as she continually tried to prove herself.

Then she stumbled upon Keith Harrell's book, *Attitude Is Everything*, and for two days, she barely put it down. She was so taken by the book's message—that your attitude will help you thrive in life—she began telling everyone around her about it, even if they didn't seem to want to hear it. For weeks, she reread the book, underlining key points and scribbling notes in the margins.

"Each page felt like my own personal therapy session and a quick reminder that I had absolutely no control over anything other than my attitude," she told me. "And that alone was enough."

After a while, she decided she needed to graduate from carrying the book around and instead, added her favorite passages to note cards. This made sense to her, because her mother used to write encouraging words on Post-it notes to remind Katie of her mother's unshakable belief in her.

She started with five or six notecards and eventually wound up with three dozen lined three-by-five cards held together by an old hair tie. Whenever she felt imposter syndrome, she pulled out her note cards.

To Katie, these were her daily affirmations, reminders that she was no imposter, and she wanted to share the good news. Even years into a successful career, she still needs them now and then, because even successful people need a good pep talk.

AUTHENTICITY MEANS BELONGING

It didn't feel like diversity of any sort was valued when I joined the military. Even after decades of desegregation efforts, the armed forces seemed to value a homogeneous group where everyone thinks the same, acts the same, and looks the same.

When you're all the same, it makes it easier to feel like you belong, and when you belong, you can act like your authentic self. I couldn't do that. Since then, the U.S. military has come to realize that diversity brings a lot to the team, with studies finding that diverse groups tend to process information more carefully, make fewer factual errors, and are more prone to innovate.[14] But in the early 2010s, the Air Force was on the cusp of a huge cultural change, and I was caught in the middle of it. I didn't think, act, or look the same as most of the people in my squadrons, and I had to quash my own authenticity to fit in, even as things slowly changed around me.

The year I spent learning to fly the F-16 felt like an initiation into a fraternity. Only part of it was about learning flying skills. Much of our time was spent being assimilated into that community. They wanted us to understand how we fit in. We got mock call signs, did roll calls, and even ate whole raw eggs, shell and all, in front of everyone else. Like I said, it was like a fraternity initiation.

By the time I got to Japan to join my first combat squadron, I was again one of two women in my unit. I was also the only lieutenant, so I was designated the unofficial squadron gopher: I fetched food, made coffee, and served popcorn. I was also the de facto lieutenant protection agency (LPA), which is a totally made up position where the young (naïve) lieutenants are tasked with providing entertainment and keeping squadron morale high. They are also encouraged to pull off spirit missions, which are essentially pranks.

Normally, you'd have three or four members of the LPA to spread out the shenanigans, like stealing other squadrons' mascots or swiping the *T* key off every computer keyboard, but it was just me. One of the first tasks that my squadron assigned to me was more serious than a prank. I had to go through everything in the cabinets of the LPA office and shred anything that could be considered inappropriate, especially relating to fighter pilot songs. That's because there'd been a complaint filed with the Inspector

General by a woman elsewhere in the Air Force who asserted that particularly raunchy songs (e.g., one was about "dead hookers") should no longer be part of the culture.

I was new to the squadron, having arrived in the country just days prior, and here I was, opening up cabinets and shredding songbooks and pornography, which pilots would take with them on long flights across the ocean, along with trivia and Battleship games to play over the radio. You know, for the sake of entertainment... I mean, it was tradition, right?

I was one of the only women in the bunch, simply following orders, shredding and thinking, *Does anyone else see the irony in this?* Watching photos of naked women get pulled into the shredder made me wonder how my peers viewed me and the other female pilot. Were we going to be taken seriously and treated equally or objectified? It was an odd thing for anyone to be doing, but for me especially, and it made me feel uncomfortable. My "welcome to the squadron" was another reminder, in case I'd forgotten, that I didn't quite belong.

How can you be yourself if you're constantly forced to ignore a big part of who you are? During my first year with my first squadron, my commander asked if I'd share my experiences as a female fighter pilot during a sexual harassment prevention training. I declined. I didn't want to stand out even more.

"I'm new here. I'm only a wingman, and I don't think people are going to respect what I have to say," I explained.

But another fighter pilot, call sign Taboo, did have something to say. A major with at least six more years of experience who was the assistant director of operations in another squadron, Taboo got up in front of everyone and told us that she'd been deployed to a base where all the briefing rooms were pretty much wallpapered with pictures of naked women. One picture after another until every square inch was covered.

As the flight lead, she was briefing an older lieutenant colonel who would be her wingman for the flight, but he couldn't make eye contact with her because the decor, which I'll call "modern *Penthouse*," made the situation uncomfortably awkward for everyone involved. Imagine you're in some conference room at work and the walls are covered in tear-outs of Miss Octobers past. Again, work hard, play hard, but this was the epitome of the work hard part of that equation. They were briefing a mission in a deployed environment where their decisions could cost others their lives.

After the flight, Taboo told the commander that the "wallpaper" was detracting from their ability to complete the mission, and the next day, they were all taken down, perhaps by another squadron's LPA. I admired her for speaking up and for sharing her story with us. It was a glimpse into what it meant to own your authenticity in a place where uniformity was most valued, and it helped the Air Force as it went through that cultural change.

WHY AUTHENTICITY MATTERS

The American Psychological Association defines "authenticity" as "a mode of being that humans can achieve by accepting the burden of freedom, choice, and responsibility and the need to construct their own values and meanings in a meaningless universe."[15] If that sounds like something you'd have to memorize for a final exam, or something that a character in a French art film might say after a long drag on a cigarette, try instead this succinct definition, attributed to Adam Grant:

"Authenticity means erasing the gap between what you firmly believe inside and what you reveal to the outside world."[16] Or as Oscar Wilde was said to have quipped: "Be yourself; everybody else is already taken." It's about presenting to the world your genuine self.

Authenticity matters because we can't fully connect if we're pretending to be someone else. When we're at our most genuine and authentic, we can create deeper relationships by earning trust. Best of all, when we act like our true selves, we don't have to operate from fear, always second-guessing how to behave.

What are the ingredients to authenticity? Psychologists Michael Kernis and Brian Goldman developed an Authenticity Inventory based on four components:

1. **Self-awareness:** an understanding and belief in your emotions, perspectives, and abilities.
2. **Behavior:** acting in ways that match your values, even if they're criticized or rejected.
3. **Unbiased processing:** being clear about your strengths and weaknesses.
4. **Relational orientation:** how you experience yourself in relation to others.[17]

It may sound simple, but it's actually difficult, because it requires us to be open and honest, even vulnerable, sharing our imperfect traits with others. But the military, at face value, isn't really set up to promote authenticity, and that would prove problematic for me during one of the toughest points in my life.

When I was in Japan, I was going through a divorce. It was very painful and lonely for me, and I was trying to hold it all together,

THE FLIPSIDE

get through the legal part, and move on. I was living on a military base halfway across the world without a support system, little work-life separation, working twelve-to-fifteen-hour days, and watching my marriage implode. I had to force myself to go to work and do my job.

I told my flight commander about the divorce, and he told me I needed to tell the squadron commander, a super-talented pilot and instructor who led the entire unit. Though he was exceptionally analytical and smart in the cockpit, he lacked emotional quotient, or EQ, skills.

Still, I expected the news of my divorce to be met with some compassion, something along the lines of "What can we do to support you?" After all, the commander regularly recited that taking care of our people was the priority. Instead, he replied, "What do we need to do to get you back to being a contributing member of this squadron?"

I had missed exactly one flight because my husband and I had an appointment with the base's military lawyer to fill out some required paperwork, and that was the only time available. But I'd found a replacement, so it wasn't as though anyone had to scramble to find someone to take my place that day. Usually, the squadron commander wouldn't even know about a minor pilot swap like that—or care.

Here, I'd shared my problem, just like my flight commander suggested, and all I got in return for my honesty and vulnerability was a request for me to become a "contributing member" again. I was shocked. It felt like salt in an open wound. That's the moment I realized I wasn't going to get any support from my squadron leadership. I was on my own in the middle of one of the lowest points of my life.

"Yes sir. It won't be an issue again."

I was struggling and some of the things that my squadron mates said and did at the time didn't help. I was so far from home and we spent so much time working that my coworkers had become my family. We even cooked Thanksgiving dinner together in our very own "Friendsgiving," and I ate slices of cake at their kids' birthday parties.

We were also extremely good at what we did. Tactically, it may have been the most effective team I have ever been a part of. Our mission revolved around the Suppression of Enemy Air Defenses, or SEAD, where we'd enter an enemy's air defense system as the frontline with the goal of getting targeted by their surface-to-air missiles, or SAMs. Once their radars found us, we could then nail down their location and, hopefully, take them out with our own weapons, clearing the way for strikers to focus on other targets without the threat of SAMs. In other words, we were bait.

The unofficial motto of SEAD squadrons is YGBSM: "You Gotta Be Shitting Me." It came from a B-52 electronic warfare officer named Jack Donovan when the first mission of its kind, serving as bait for a North Vietnam SAM site, was proposed in 1965. He was incredulous, saying he had to "shoot it before it shoots me...you gotta be shitting me!"[18]

SEAD missions are very technical and complex, and require a large amount of coordination and trust, which bonded us pretty tight. So it feels like I'm betraying them now by saying something negative about our time together. It was an honor to be part of such an effective team, but at the same time I was doubting that I belonged in it. I wonder now how much more amazing that assignment could have been with the same people, the same mission, but without the sense of isolation.

Words we use now, like "microaggressions," "misogyny," and "toxic culture" carry a lot of power, but they had different meanings then, if any at all. I don't think any one person in that squadron had the intention to isolate me or make me feel unwelcome. I doubt they realized that halting conversations when I walked into a room, making degrading remarks about women in general, or waiting for my reaction to gauge what was socially acceptable slowly eroded my ability to feel like part of the team. They embraced the cultural norms of a fighter squadron at the time, and so did I. We've all since grown and changed, and we aren't now who we were then. Many of them would likely cringe at behavior we thought was normal at the time.

Now I see that it kept me from being my authentic self. I had to play dress-up to endure their "othering" of me, and that blocked me from a very crucial outcome of authenticity: vulnerability. When authenticity is cordoned off, there's no room to share our emotions, beliefs, and thoughts. When authenticity is quashed, so, too, is vulnerability.

DO VULNERABILITIES GET YOU SHOT DOWN?

You might think that being vulnerable would be considered a bad thing in the military and in many ways, you're right. In combat, being vulnerable means being susceptible to attack. That's why we check our six (see chapter eleven). Here, vulnerability is dangerous.

But I'm not talking about physical vulnerability or military tactics. I'm referring to emotional vulnerability, which Brené Brown defines as "uncertainty, risk, and emotional exposure." Why would anybody want to endure that? Brown says, "Staying vulnerable is a risk we have to take if we want to experience connection." In her book *Daring Greatly*, she asserts that vulnerability isn't a sign of weakness, but instead, "the birthplace of love."[19]

When it comes to leading a squadron of fighter pilots, or any soldier, what's love got to do with it? When you lead with vulnerability, you lead with love and that brings the best out of everyone. Justin Maciejewski, a former commander of the British Army's Second Battalion, the Rifles, explained in an interview with *McKinsey Quarterly*, a business publication, how vulnerability worked during peacekeeping missions in Basra, Iraq, in the 2000s.[20] He'd show up at the guard tower at 2:00 a.m. and ask a young soldier, "How are you feeling?" Then they'd talk about everything they were both concerned about and how to address those concerns. That's vulnerability in leadership, and it matters because one of a leader's primary jobs is to open the lines of communication. Without it, people can feel hesitant to try the kinds of things that help them—and their teams—grow.

I have been on the receiving end of leaders who were vulnerable. They're the ones I felt I could share things with, while the others I couldn't. My boss in Japan obviously made it clear I couldn't. In sharp contrast, my boss in Texas, where I was stationed after Misawa, was very vulnerable with me about his leadership struggles or the personnel dilemmas he had with various people he had on his large team. It was clear to me that he really cared about his team, so it was a lot for him to carry when they had hard things happening in their lives. When he admitted he wasn't unflappable, it opened the door for us to trust him when we needed to share something vulnerable.

Think of it this way: If you feel that you have to be something else and act like someone else, that you have to pretend, you'll likely spend your days in a hypervigilant state, second-guessing your every move and word. This desire to be perfect keeps you from trying new things, which prevents innovation. Meanwhile, vulnerability from the top down makes room for everyone to try and fail and try again because it is positively modeled for us. This happened to me with another boss, on the Thunderbirds, who appeared to have it all figured out until he finally shared something deeper. Most days, he would share minor, surface problems, like "'I'm tired today," but nothing of depth, so I rarely shared anything deep with him, either. Then one day, he overheard me say that it was hard to be an introvert and a Thunderbird, which required us to make public appearances and interact with the crowd. Sometimes, it drained my energy level.

"Mace, this isn't natural for me, either," he admitted. "I am introverted, too." He didn't have a solution, but I didn't need one because his vulnerability allowed me to identify and bond with him for the first time. It allowed me to see him out there interacting with the crowd day in and day out as the leader of the team and realize that he wasn't perfect. This one remark shifted our dynamic for the better. Being vulnerable as a leader doesn't mean crying to your team or airing your dirty laundry. The gestures can be small and simple.

In a video called, "Let's Talk About Vulnerability," Simon Sinek explained that he was unsure how to lead when the Covid-19 pandemic hit.[21] He decided that if he faked happiness, it would have put pressure on everyone on his team who was struggling to do the same. High performers can do that, he said, doubling down on work during stressful times. But that's a mistake.

"The ability to be vulnerable is one of the greatest and most difficult leadership lessons any of us can learn," he explained. "But it is so important to building trust when people see us as human because it allows them to be human, too."[22]

We tend to fear failure, but it's a fear of vulnerability that leads to missed opportunities and limits our potential. We don't like feeling exposed. Imagine that time you had an idea in a meeting but were too intimidated to raise your hand. Or you saw that cute guy or girl across the room, but never had the courage to introduce yourself. Putting yourself out there is scary. But I promise you, the cost of stagnating because of fear outweighs the discomfort that comes with vulnerability.

Scriptside:

- Letting imposter syndrome stunt your growth.
- Pretending you know more than you do.
- Hiding your true self.
- Letting fear of vulnerability get in the way of opportunities.

Flipside:

- Recognizing the normalcy and cause of imposter syndrome.
- Acknowledging your beginner's mindset and beating the naysayers to your inconvenient truth.
- Leaning into your authenticity.
- Making vulnerability work for you.

CHAPTER FOUR
Remember to Wiggle Your Toes

I'm a brand-new flight lead, guiding four jets through an air combat training exercise involving some fifty aircraft of various kinds: F-16s, F-22s, F-15s, bombers, cargo planes, surveillance aircraft, and refueling tankers. All those planes have their own tasks that support the mission, so it's a lot of coordination. Some of us have to go in and take out the surface-to-air missiles while others "eliminate" the bad guy jets in a complex simulation exercise. With so many aircraft airborne, it takes a lot of awareness not to hit each other, much less for each moving piece to go entirely as planned.

My flight's initial mission is "air-to-air"—targeting enemy jets to keep them from attacking our own. The four of us start pushing into bad-guy territory and their aggressor aircraft point back at us. It's already pretty stressful, but then one of my most experienced pilots, my second-in-command, gets "killed" right away in simulated fire. So now it's just me and two wingmen.

We eventually need gas, which means we have to circle back to the tanker that is hanging out behind the battle lines. Only, there's aircraft *everywhere*, making it hard to locate our assigned air refueling plane. Plus, I have several jets flying close formation on my wings. I can't just "roll and pull," turning sideways and then flying "up," to put us behind the tanker. Every angle and turn needs to be gentle and thought through for us to successfully "park" our formation just behind the other aircraft.

When we finally reach the tanker, I get to refuel first because I'm the flight lead. I have to get onto the boom, the long tube that extends from the tanker aircraft holding the fuel, get my gas, and get off. Oh, and my gas tank's opening is behind the cockpit and therefore, behind my line of sight. Thanks, engineers! Even moving only my head causes tiny inputs on the F-16 stick that can lead to big movements of the jet itself. As a result, twisting around to watch the boom approach the air refueling port is not an option.

Not only are my wingmen watching me while waiting for their turn, there's also a four-ship of Raptor pilots—F-22s—queuing up, and we're all using the same radio frequency, so they can hear what's going on. I'm a little nervous, and with good reason: I'm in the wrong spot. I try to position my jet under the boom, but my airspeed is off and I overshoot it, so I have to fly back to the starting position. I mumble curse words under my breath and get ready to give it another try.

A delay in one aircraft hooking up can quickly put others in a position where they are running low on fuel and may need to return to base, leaving the fight. My inability to quickly and skillfully put my aircraft in position could be the thing that leads to mission failure for the entire team.

For someone already struggling with imposter syndrome and a need to prove myself, this feels like a test, and one that I dread. I've air-refueled only a couple times, and that was months earlier. I am by no means proficient at this complex skill yet, and it shows. But this is no time to panic.

Air refueling is even more difficult at nighttime or in inclement weather, like inside a cloud. This lack of depth perception and inability to see the horizon to assist with spatial orientation can do weird things to your body. Spatial disorientation, or spatial-d, occurs when there's a mismatch between what we sense visually and what we feel due to our inner ear.

Anyone who has experienced vertigo knows just how confusing this can be. It's not uncommon while air refueling to be certain that you are flying straight with your wings level, only to glance down at your attitude indicator to discover that you're actually in a thirty-degree banked turn.

I would go on to experience this many times, and the mental fight to talk your brain into believing your instruments while keeping the jet in the right position is one of the more difficult things to manage in the cockpit. There are plenty of stories of pilots unintentionally doing barrel rolls around a tanker at night as they attempted to get what they were seeing and what they were feeling to match.

Under all that pressure, it's easy to stress out, and soon the jet's all over the place. New pilots in particular are at risk of putting the jet into a pilot-induced oscillation, or—because the military loves a good acronym—a PIO. It's when a pilot takes an increasing series of corrections in opposite directions until they cause the jet to pitch and wobble up and down until it's "porpoising."

My flying career would have more overcorrections throughout the years. This wouldn't be a huge deal when I was by myself, but when I was trying to attach to a boom midair or fly as close as eighteen inches from another Thunderbird jet, the stress levels and potential repercussions were high. Flying in very close formation in Thunderbird training and performances required maintaining a steady jet like I'd never had to before. In combat, unless the weather was so bad we had to stay close enough to see one another inside a cloud, we rarely flew that close. Plus, Thunderbird flights frequently weren't level or right side up, but rather upside down or sideways. Overcorrect in that situation, and in a split second, you could be dangerously close to the aircraft next to you or worse.

On this particular refueling flight, my heart rate is rising and my hands are on the verge of cramping on the stick as I attempt the tiniest inputs to position my F-16 to the boom. But the harder I try to control the aircraft, the more my arm feels like a cement block, and the more frustrated I become.

So I do what my instructor pilot, call sign Siren, once told me to do: I wiggle my toes. I always looked up to her, but frankly, "wiggle your toes" sounded silly to me the first time I heard it. *I'm flying hundreds of miles per hour in a $20 million aircraft*, I thought, *and she wants me to wiggle my toes?* Yet this simple set of directions completely changes how I approach the boom and many other stressful missions to come.

Siren said that when you start to feel like you're overgripping the stick and your hand is starting to go numb, think about your toes inside your boots and focus on wiggling them. It makes you relax and stop freaking out about all the things that might go wrong in the air, making it the perfect tool to help me avoid a PIO and connect to the boom.

Siren would go on to teach me how to be a better fighter pilot and how to navigate my new role in the community. Later, she'd become a friend and a mentor. Yet one of the greatest lessons she ever taught me was to wiggle my toes because it helped me to snap out of my hypervigilance and refocus on the right things.

THE FLIPSIDE

If this sounds like I'm about to tout mindfulness and suggest that you sit in lotus position on a mountaintop, that's not where I'm going with this. Not really. Not that mindfulness doesn't have its place, as a study of American military found that mindfulness training had a positive effect on working memory capacity, which is used to regulate emotions and manage cognitive demands.[1]

But wiggling your toes is something different from mindfulness, and it's very much not meditation. Flying an F-16 is not the time to observe my thoughts passing by without judgment. Rather, it's about refocusing. The Air Force used to advise pilots when something went wrong in the air to "wind the clock," meaning they should reach down to the old-school clock positioned in the cockpit by the pilot's right knee and wind it. It was a physical disrupter of panic, a refocusing of attention, much like wiggling your toes, that disrupted the panic of a pilot rushing to make a flash decision in a situation that might need more analysis.

What you need to avoid your own version of a PIO is focus. Christina Bengtsson, a Swedish military officer, fighter aircraft technician, and precision-shooting world champion, describes focus as "a mindset that enables you to take concrete action toward your goals."[2]

In her TED talk, she said humans struggle with focus for three reasons: (1) Their minds are often full of thoughts of worry and of not being good enough. (2) We tend to focus on the future instead of working with what we already know. (3) We are frustrated by time constraints.[3]

In a PIO, you hit all three. We're worried we can't get on that boom or into that formation and that everyone's watching us try. We can get stuck on how we don't have much experience in midair refueling or flying in formation instead of thinking about how we're among the pilots who are proficient enough to even try. And we feel the pressure to get really good at it right away.

I bet it's the same for whatever your version of a PIO is. You're anxious or worried, so your thoughts are oscillating back and forth between *I can do this* and *I'm not good enough*. You're thinking about what you don't have now or haven't achieved yet, and you feel like you're running out of time. It doesn't have to be that way.

ARE YOU OPERATING IN THE RIGHT BRAIN WAVES?

I met Stefanie Faye, a neuroscience specialist and educator, while attending a retreat with Veterans' Outdoor Advocacy Group in Colorado. She was there to teach the veterans—a group of ex–fighter pilots, former Green Berets, and retired Navy SEALs, many who now run nonprofits focused on improving mental health—how our brains operate. When I told her about Siren's advice to wiggle my toes, she confirmed that some of what we know intuitively or from experience is backed by science.

"We are wired for evading life-threatening danger," she explained, adding that we often choose what appears to be the quickest way out of tough situations. "You're not going to contemplate multiple possibilities. You're going to go for the fastest route possible." Wiggling your toes, she said, takes the focus off the immediate cause-and-effect, and shifts your brain into the medial prefrontal cortex, where you're able to observe yourself.

"It's basically diffusing that beam of awareness to not be so tightly gripped on what you think is the problem." She says that when I shifted my attention from clutching the stick to wiggling my toes, I stopped being hyperfocused, diffused my energy, and expanded my awareness.

It's likely my brain had been in high beta, the so-called emergency brain wave. Brain waves are patterns of electrical activity, and though they're difficult to measure in controlled studies, Stefanie says she has witnessed experientially how they affect people.

Beta is typically our waking state, a general alertness that is correlated to fast thinking. High beta brain waves are the fastest waves in this state.

"They're generally related to a clustering of neurons firing," Stefanie said. "The closer together they are, the faster the frequency."

To get a sense of where beta waves fall, take a look at the most commonly measured brain waves:[4]

> **Delta (1 to 4 hertz):** The slowest waves, which occur mostly during deep sleep or unconsciousness.
> **Theta (4 to 7 hertz):** Occurs in deep meditation and the earliest stages of sleep, like when you're just drifting off.
> **Alpha (7 to 12 hertz):** Occurs when we're awake but in a resting state.
> **Beta (12 to 40 hertz):** These are faster brain waves that are typically engaged when we're solving complex problems or using logic. Too much of it can lead to anxiety.[5]
> **Gamma (40 to 70 hertz):** These superfast brain waves are involved in processing information and operating at peak concentration, a high alert.[6]

"People who are very anxious, people who ruminate a lot, who have OCD tendencies, often have a higher kind of volume in high beta," Stefanie said. She explained that shifting your beam of awareness into a lower, slower brain wave can help you regulate even in high-pressure situations. Wiggling my toes allowed me to open my beam of awareness beyond my hand gripping the stick and potentially moved me into a more useful brain wave. Stephanie describes it as a way of "anchoring back into our body" because we're noticing what's happening internally instead of staying focused on the outside stressor.

Stefanie also shared the importance of regulatory flexibility, or the ability to regulate your emotions depending on context or situation. One of the components of this skill is context sensitivity, which she defines as using appropriate and adequate energy and resources—not acting urgently when the situation isn't life-threatening and not being too chill when you're doing something that requires focus. Did your dinner just catch fire on the stove? You're going to need high beta to stay hyperalert enough to deal with that emergency right away. But if, let's say, your mashed potatoes came out chunky and your dinner guests just rang the doorbell, you don't really need high beta to deal with it, do you?

Another part of regulatory flexibility is what's called the "repertoire" of strategies for navigating various situations. In other words, we may have different coping mechanisms for dealing with problems and situations. "What they have found in research on resilient individuals," Stefanie said, "is that it's extremely difficult to predict which strategy they're going to use because the most resilient individuals are the most flexible."

Wiggling your toes is one tool you can use to notice things other than what's freaking you out, and that, Stefanie said, is an important strategy for resilience. And though it seemed at first like silly advice when I was flying a fighter jet, I came to understand that it's a helpful hack for ultimately bouncing back in tough situations.

THE FLIPSIDE

ASSEMBLE YOUR DREAM TEAM

Wiggling your toes can also interrupt the inner critic's trash talk, providing you with a moment to exercise your regulatory flexibility and perhaps say something to yourself that's more useful than "Loser!" I'm sure someone's told you to consider what your best friend would say and then to tell yourself that instead of the garbage your inner critic has tossed at you, but I think it's beneficial to draw on the wisdom of more than just one person in stressful times. You need an entire team, a Dream Team. These are people you choose to replace your inner critic inside your head. You're not calling them up in real life and asking them for their input, especially considering not all of them will be in your inner circle, or alive for that matter. Or real. Instead, you're *imagining* how, say, your fifth-grade teacher, your dad, Serena Williams, and Batman might advise you. Instead of rushing to a conclusion all on your own, you're picturing other people serving as advisors or muses, providing guidance or inspiration on how to excel in each part of your life.

Your Dream Team is personalized to you and customized to the situation, which means you can make them anyone you want. Each person you select should be especially equipped to guide you in a specific part of your life. And you can pick the best part of them, because all of these people are likely flawed in various ways, but you are choosing to embody the most admirable and applicable part of them.

Let's say, for example, that you're an athlete who loves weightlifting, and you know that Dwayne "the Rock" Johnson is recognized for his intensity in the weight room. If you're debating skipping your last set for the day, ask yourself what the Rock would do. He fills the role of your fitness muse on your Dream Team.

But your Dream Team doesn't just stand by for advice on actions. It also helps you with emotions. Perhaps you struggle with anger, blowing up a little too easily, especially at home. Who can you turn to? Maybe your grandparents, who have been married for fifty years and always treat each other in a loving and respectful way that you'd like to emulate. Whenever you feel that anger start to rise, wiggle your toes, and choose to channel Meemaw instead. She's your muse when it comes to being a loving partner, inspiring you to respond in a healthier way.

Your Dream Team represents your own inner wisdom, a sort of anti–inner critic. Though they are parts of you, assigning them identities separate from yourself makes it easier to muster the courage that you see in others but struggle to access within yourself. If you're pretending to be someone who's more confident, more fearless, or more loving than you've yet learned to be, it can help you handle stress and anxiety better. Reframing is possible because the physiological response to stress and anxiety is the very same one we get when we're excited. No wonder: They all hang out in high beta.

That's why author and speaker Simon Sinek says out loud to himself, "I am excited," when his nerves threaten to get the best of him. He discovered this trick while watching the London Olympics when, he says, every interviewer asked every athlete, "Are you nervous?" or "Were you nervous?" Yet every athlete responded, "No, excited."[7]

He figured that world-class athletes must surely understand something important about body stimuli. He channeled them when he was about to speak to an audience of three thousand chiefs of police during a wave of police brutality after he noticed his palms were sweaty and his heart was racing. He even visualized the outcome of the event, just like Olympic athletes do. The stakes were high, but he knew that both feelings, nervous and excited, trigger the same stimuli. He would use that to his advantage.

He wanted to change his narrative, so when he was backstage before the event, he said out loud, "I am excited!" Soon, he said, he truly felt excited to speak to the audience, because they had the power to make changes at a crucial time in history. He says it made him want to rush forward instead of pull back.[8]

Was I nervous to be a part of the flyover during the national anthem at Super Bowl LIII in Atlanta in 2019? You bet. I'd joined the Thunderbirds only a few months earlier, and this would be my first event in front of (or over) the public. I was thinking about how 100 million people would be watching the flyover at the tail end of Gladys Knight's rendition of the national anthem. I'd be on TVs across America and on the Jumbotron in the stadium.

That's a huge stage to fail spectacularly on. I was nervous. I was also excited.

The Super Bowl fell about three-quarters of the way through our winter training season, the time when there aren't air shows and new pilots are learning to be Thunderbirds; I was one of those newbies with just three months of this kind of flying under my belt. After a few months of flying in separate diamond and solo formations, we were going to fly delta, when six jets form a triangle, kind of like birds flying south for the winter. I was flying on the far outside left wing and I had to make sure I kept my F-16 close to the others when they turned right just after we passed over the stadium, or else I'd get "hubcapped," left behind by the formation. In this case, it would have been in front of an audience the size of about a third of the entire American population.

I didn't get hubcapped, but I didn't exactly hold tight to formation, either. I drifted a little bit to the left during the turn, and that was the thing that stuck with me well after it was over. That disappointment kept me from fully enjoying the incredible experience I was having.

After we landed from the flyover, we had a police escort to the stadium, where we hung out in the tunnel by the locker rooms and watched the players from the Rams and the Patriots coming in and out. I mean, we saw Tom Brady stretching out. On the field, Adam Levine and Maroon 5 were performing the halftime show.

Eventually, the Thunderbird team was announced, and we headed out onto the field, where a stadium of nearly seventy thousand people gave us a standing ovation. At the end of the game, we got to stand on the field again as confetti was falling so thick, I would find pieces of it stuck in my showsuit weeks later. What an insane, once-in-a-lifetime moment.

And what was I thinking about right then? How I messed up formation in the flyover.

It wasn't like I could dismiss it and say, "Nobody saw that." It's the *Super Bowl*. Though I knew that the general public wouldn't notice my mistake, other fighter pilots and Thunderbird alumni would, and that bothered me.

Here was an opportunity for me to reframe it and let go of what I couldn't control. I had just had the experience of a lifetime, and I was choosing to miss out on the joy that was part of it. The flight had already happened, so why let a minor mistake poison this incredible experience? And you know what? Nobody said anything about it to me. Not once. I was the one who brought it up with the Thunderbirds' former Left Solo, who'd been in my position before me, and he admitted he'd seen it. But he had sympathy for me, because something similar had happened to him in a Daytona 500 flyover. The formation turned right and he got left behind. Hubcapped.

If I'd had a Dream Team for the Super Bowl flyover, I might have chosen Tom Brady to be on the list. He was surely under much more scrutiny that night than I was. But on my Dream Team, it's his experience under pressure I'd have needed. If he got caught up in his mistakes or lost opportunities, dwelling on what's gone wrong, he wouldn't be able to focus on the next play, the next throw, or even able to celebrate the next touchdown.

I might have also called upon another fighter pilot from my first assignment that I looked up to, not only as a tactical expert, but also as someone who handled mistakes with grace and acceptance. Finally, everyone needs a cheerleader on game day, so I would have pulled my mom off the bench. What a motley crew my Dream Team would have been, but they would have helped me keep my mistake in perspective so I could embrace the positives.

STOP FIGHTING THE JET

When fighter pilots get into a PIO or other troublesome situation, it's often because of what we call "fighting the jet." When we fight the jet, we're trying entirely too hard to control something that doesn't need that much controlling. And when you fight the jet, you will almost always lose.

Aircraft are inherently stable. It's actually fairly difficult to put the F-16 out of control because of all of its finely tuned computers. That's true of most fighter jets. In fact, a pilotless F-106A actually *landed itself* in a training mission in 1970. The jet had been in an uncontrollable flat spin, forcing the pilot to eject near Malmstrom Air Force Base in Montana. With the pilot out of its cockpit, the jet somehow righted itself and made a belly landing in a snowy field, earning its nickname, the "Cornfield Bomber."[9]

More recently, an F-35B fighter jet kept on flying after its pilot, a U.S. Marine, ejected, traveling for sixty miles before crashing.[10] It flew so far that the Marines had to ask the public to help find it.[11]

Whatever situation you're trying to overcontrol by fighting the jet is probably more stable than you believe. The offspring who left her lunch on the kitchen counter this morning will figure out how to get something to eat during lunch period today. Home buyers are not going to notice that you drove to every Target in the county to find matching hand towels for the bathrooms. Your PowerPoint presentation is good enough, so leave it alone.

"But...but...but you don't understand..."

Before you try to argue, did you know that the very first step to take charge of an out-of-control fighter jet in a Critical Action Procedure, the guidelines pilots follow in emergencies, is controls *release*? Because the jet will usually recover itself, perhaps even belly-land in a field. Let's face it, you're probably the one who put the jet in the situation that made it go out of control in the first place, so why keep on doing what hasn't worked?

The next time you're in a spiral of your own making, release the controls! If you have to sing like Elsa from *Frozen* to "let it go," then do that. Or imagine it as step one on your own Critical Action Procedure checklist, because if you keep fighting the jet, it has zero chance of righting itself.

Chances are, you've done enough work already that you don't need to attempt to overcontrol. Lunch was left on the counter?

You've taught your kids how to ask a teacher for help. Want someone to make an offer on your house? Focus on the things buyers actually look at and let the rest go. Hoping to make a kick-ass presentation? Check for typos and do a run-through, then go get a snack and stop obsessing. When you stop fighting the jet, you release fear and start to think more clearly.

KEEP CALM AND CARRY ON

In 1939, the British Ministry of Information created three posters designed to assuage the anxieties of Great Britain's public at the start of World War II.[12] The government thought the country would soon be subjected to a sizable aerial bombardment, "the Blitz," and they wanted to raise morale among an understandably nervous public.

The first two posters, designed in all caps, white type on red ink with a drawing of the king's crown at the top, were released and posted around the country. They read:

- Your Courage, Your Cheerfulness, Your Resolution Will Bring Us Victory
- Freedom Is in Peril Defend It with All Your Might

These two posters didn't go over too well. Imperial War Museums curator Claire Brenard explained that the posters failed to instruct the public on tangible things to do at a time of war.[13] Plus, they weren't snappy enough to be memorable. The Blitz didn't even happen for another year, and the posters were quietly retired.

The mindset the posters hope to encourage was one of endurance through some of the toughest of times the United Kingdom had seen. When the bombs eventually fell from the sky, Brits were supposed to understand that they needed courage, resolution, "cheerfulness" (aka the British stiff upper lip), resilience, and some fight to get the nation through it.

But there was a third, more clever poster that had very limited distribution:

- Keep Calm and Carry On

If that sounds familiar, it's because, thanks to the internet, this WWII poster had a resurgence in the twenty-first century. It was first mass-produced during the financial crisis of 2008, more than sixty years after it was designed, and was soon made popular thanks to Instagram and other social media platforms.[14] Of course, online sellers on sites like Etsy began offering customized posters (and T-shirts, mugs, beer cozies, and more), and as a result, now we can buy, "Keep Calm and Crochet On," "Keep Calm and Call Mom," and "I'm Spanish. I Cannot Keep Calm."

A bookstore owner in northern England had found one of the original WWII posters at the bottom of a box of books and made copies and sold them. In fact, he's sold thousands of copies. When asked why it was so popular, he said, "It's cheaper than antidepressants."[15]

"Keep calm and carry on" is a mindset, and apparently it resonates with today's generation even though it was designed for our grandparents and great-grandparents. That says a lot about our problems as humans: The details may change, but at the core, we're all struggling with the same things.

THE FLIPSIDE

"Keep calm and carry on" could easily fit with "Wiggle your toes" and "Don't fight the jet." To pull off the keep-calm part, you need to first wiggle your toes to take your focus off the issue at hand, and then release the controls so you stop doing what's not working and quit fighting the jet. Assembling your Dream Team is the start of carrying on.

All these concepts help you release the control that your inner critic has stolen from you. It has stormed your inner peace and left you with fear, doubt, anxiety, and many other things that don't suit your needs. Once you begin to master keeping calm, your inner critic will have less and less to say. That's when you can begin to carry on.

Scriptside:

- Creating a PIO by overcontrolling.
- Remaining hypervigilant.
- Going it alone.
- Fighting the jet.

Flipside:

- Wiggling your toes.
- Building your resilience.
- Assembling your Dream Team.
- Releasing the controls.

CHAPTER FIVE
Take One Minute, One Hour, One Month for Change

I was pushing up my jet's throttle, watching my airspeed rapidly tick up, and assessing the distance that was quickly closing between my jet and the other Thunderbird solo pilot's aircraft. I was chasing him down so that the two of us could join up with the remaining four aircraft and create the iconic six-jet Delta formation. It was day one of our performance in Colombia, the first time the Thunderbirds had performed in that country in fifty years. Below, the jungle zoomed by, a blur of green as my jet reached four hundred miles per hour, more than one hundred miles per hour faster than the other aircraft that I was gaining on. Suddenly, I saw a black flash under my jet's nose and felt the aircraft shudder.

It was a bird. A very large bird. The GoPro camera mounted inside my aircraft's canopy captured video that would later reveal it was likely a black vulture, which weighs up to six pounds and has a wingspan of nearly six feet, about the length of the average couch. They can fly up to thirty-seven miles per hour—which, sadly for the bird, wasn't fast enough that day. Think of the bird strike on

Captain "Sully" Sullenberger's airplane that led to an emergency landing on the Hudson River, with a bird that looked to me as huge as a pterodactyl and a jet flying more than twice the speed of the average commercial airliner.

I didn't know if my engine was damaged or if some other critical system had been destroyed by the impact I'd felt. I needed to land as soon as possible. I didn't know if I'd need to burn down fuel so the jet would be lighter to land or if my landing gear had been damaged. There were many unknowns, and I was still flying over the Colombian forest at more than three times the speed of the world's fastest roller coaster.

My adrenaline spiked, my heart raced, and my mouth went dry. Then, without even consciously thinking it through, I fell back on my training for prioritizing in an emergency:

1. Maintain aircraft control.
2. Analyze the situation.
3. Take proper action.
4. Land as soon as conditions permit.

Would I end up piloting a twenty-thousand-pound, engineless glider that wasn't very good at gliding? Would I have to eject over the Colombian jungle, sending a $20 million plane—and me—into the trees? I scanned the instruments: oil pressure, engine temperature, RPM, hydraulics. They all held steady. I sniffed the air inside my mask, because if a bird had ended up in the plane's intake, there'd be a burnt chicken smell in the cockpit, or so I'd always been told. I'd hit birds before, but nothing like this.

"Six needs to terminate," I radioed to the other five pilots in the middle of their own air show maneuvers. It meant that something was wrong, and we had to pause the show to assess what to do. I needed to land the jet, and right away. When I touched down on the tarmac and taxied to park, my dedicated crew chief, one of the enlisted maintenance professionals and a guy who's usually pretty calm, climbed the ladder to my cockpit, wide-eyed.

"Ma'am? You have two huge holes in the side of your jet."

REACT VS. RESPOND

I landed safely on the ground because I responded instead of reacting. When you *react*, you're providing an immediate and instinctive action, but when you *respond*, you've taken a more deliberate and thoughtful approach.

Thousands of years ago, we had to react to a rumbling herd of animals running in our direction when we lived out on the Serengeti. But back then, we didn't have the luxury of responding. Now, we usually do, even if we still have to stay vigilant when it comes to our safety.

University of San Francisco professor Jim Taylor, PhD, says the difference between "reacting" and "responding" isn't just semantics, and that difference is exacerbated by the demands of modern life. "Reacting based on our primitive instincts or baggage rarely leads to positive outcomes," he wrote in *Psychology Today*.[1] He says that we don't need to rely as much on our amygdala, the brain's threat detector. It's what guided us to avoid being eaten by tigers 250,000 years ago.

Relying heavily on the *react* part of the brain can be troublesome. In *The Obstacle Is the Way*, author Ryan Holiday writes, "A deer's brain tells it to run because things are bad. It runs.

Sometimes, right into traffic." He says it takes strength to override what he calls the "animalistic brain." It takes practice.

That was what I was doing right after the bird strike: letting the animalistic brain react without getting stuck there. It processed the sudden *BOOM* sound coming from under my jet, but it was the prefrontal cortex, which provides executive functioning—analysis, planning, and problem-solving—that helped me land the plane. My executive functioning took over and responded.

When you're flying hundreds of miles per hour over the jungle, staying in react mode can lead to disaster. Or as social worker Shari Simmons explained in her TEDx talk, "When we stress, we regress." And where do we regress? She says into the limbic system, the survival center of the brain which houses the amygdala and reacts in one of six ways she's identified: fight, flight, freeze, faint, fornicate, or feed.

You may have heard of the flight, fight, and freeze parts—run, fight back, or, well, freeze—but not the others. She describes "faint" as checking out completely, "fornicate" as things like sex addictions (though that would be quite the choice while flying a jet), and "feed" as eating too much or too little. She warns that we as a society have become so numbed out on adrenaline, which fuels these six *F*s, that we've forgotten how to perform the simple process of assessing situations in the prefrontal cortex. We've forgotten how to respond when that's what we really need most of the time.

FLIPPING TO TRANSFORMATION

How do you move from the reaction that doesn't serve you to a more intentional and productive response? If you've just had your own form of a bird strike and you're reacting, give yourself some time, but not too much. I use this guideline:

> ONE MINUTE TO PAUSE.
> ONE HOUR TO ANALYZE AND REFLECT.
> ONE MONTH TO TRANSFORM.

Of course, the amount of time can be adjusted, but you really shouldn't let your pause last much more than a minute, or your analysis and reflection to spill into another hour or more, or your transformation to take months or, let's face it, years. Sometimes, speed matters, especially if you're flying a jet, driving a car, or keeping a toddler from dashing out of Target toward the parking lot. Split seconds, tops. But in other situations, the pause can be longer.

Mentality coach Sabastian Enges says the difference between high performers and everyone else is that they respond much faster. He says, "An average guy will have a bad day or a bad week. A high performer will have a bad fifteen minutes." It takes practice to get to that level, and it's important to point out that it's not about stuffing down your emotions.

The pause has power. It activates the parasympathetic nervous system, the counterbalance to the sympathetic nervous system where the fight, flight, freeze, and other F parts (yes, that one, too) happen. Your sympathetic nervous system revs up your heart and breathing rates, putting you on high alert, and your parasympathetic nervous system returns your body to its "rest and digest" default setting. The pause gives you a moment to recognize if you're having an emotional knee-jerk reaction that doesn't serve you.

In our fighter jets, we didn't have the luxury of pausing for more than a millisecond. In fact, when I sneezed in the cockpit while flying in a tight formation, I tried to keep my eyes open because I had to be ready to make an adjustment with no delay. (Despite what the internet might say, sneezing with your eyes open doesn't pop your eyeballs out of your head.)

PRIORITIZING THE PILOT'S WAY

Lucky for you, you're not traveling four hundred miles per hour, and it's likely that you can take the full minute, if not a little bit longer. Yet you can apply a fighter pilot's emergency prioritization system to determine your next steps. After the bird strike, I used it to make a safe landing:

Maintain "Aircraft" Control: I know, I know. I just told you in chapter four that step one is to *release* controls, and now I'm telling you to *maintain* control. But a PIO is caused by the pilot's inputs, and a bird strike is an emergency of someone else's making. (RIP, bird.)

In layman's terms, maintaining control is pretty much keeping your shit together. Fighter pilots have to, because if, for instance, we let a flashing warning light on the dash distract us, we could wind up flying into the side of a mountain. These sorts of crashes have happened plenty of times. We misprioritized our attention to a disastrous outcome.

Maintaining control is as much about the pilot as it is about the jet because if we lose control of ourselves, we can't control the aircraft. As a result, we learn to respond instead of reacting in a panic. Though "respond" and "react" are synonyms for each other, they have different meanings in the face of adversity and struggle, and most of the time, responding will be the better choice. For instance, if you get a bad review at work, don't run off and quit, bad-mouth your employer on social media, or get a headhunter on the phone. Maintain control and take the next step.

When I witnessed a friend and fellow officer receive incredibly harsh feedback from her boss, I wondered how she was going to react—or respond. Harsh feedback like that happens in the military, but this came with no warning and no prior indication that she wasn't meeting his expectations during a particularly tough assignment. At the end of the conversation, he even threatened to fire her.

She—let's call her Jen—was blindsided and devastated. When you're used to performing well, something like this can shake your entire identity, and most of us want to do our best, to not let others down, and to get positive feedback. That evening, I sat with her, and I listened and helped her run through her options. After giving herself the grace to feel the disappointment and the shame, she resolved to show up the next day and prove her boss wrong. In the coming weeks, she kept it together at work and resolved not to lose her cool.

Analyze the Situation: In the case of my bird strike there were so many things that could have gone wrong depending on where the aircraft had been struck: the engine, my landing gear, a wing. It was up to me to Analyze the Situation and then decide how to respond.

It may feel like you've been blindsided, but you often have more time to Analyze the Situation than you think you do, which is why I suggest we take one hour to analyze and reflect.

Jen could have reacted immediately and apparently; that was what her boss had expected. He was so prepared for her to quit, he even had a replacement on standby to fill her position. Instead, she gave herself the grace to feel the disappointment and analyze the situation.

You, too, can pump the brakes and pause first, perhaps consult a wingman or summon your Dream Team and consider how they'd respond. No matter what you do, a pause is in order.

It works for both smaller problems and life-changing ones, like a bad diagnosis or a divorce. Gather information, ask questions, and, like they say in Alcoholics Anonymous, "do the next right thing." Though the AA recovery community has been credited with coining that phrase, it actually comes from famed psychiatrist Carl Jung, who advised people to "but quietly do the next and most necessary thing." I say you should take some time to figure out exactly what that thing is if you have the luxury to do so.

This is where grace meets discipline, intentionally pausing in the moment to take time to seek out more information and get your initial reaction under control. But don't get stuck there or you'll find yourself dwelling on problems and spinning your wheels over issues that aren't serving you. That's analysis paralysis. Look, bad stuff happens, and it's not healthy or realistic to compartmentalize it completely. Those contracts with yourself that I mentioned in chapter one need to feel like they push you, but they also need to have a bit of wiggle room, because life happens. But soon enough, it's time to take action.

Take Proper Action: After Analyzing the Situation, hopefully in around an hour, it's time to Take Proper Action by responding, not reacting, before you can get back on the ground.

No matter whether what you're handling is a minor blip or a life-changing obstacle, some things are in your control. Sometimes it's obvious what your role in causing the situation is, and sometimes taking ownership means realizing that you have the ability to choose your response to the hand you've been dealt.

After Jen took the time to Analyze the Situation after that

less-than-stellar performance review, she started to make meaningful changes. After all, she was in a tough spot with a lot of responsibility, and everyone was expecting her to do her job.

In time, she executed changes based on every piece of feedback her boss had given her, and she did it with resolve. It was an incredible feat of strength to be there for her people even though they didn't know about the behind-the-scenes conflict with her boss and within herself. She was able to make decisions more calmly and intentionally—responding instead of reacting—thereby saving her job.

Land as Soon as Conditions Permit: Now it's time to "land the plane." You've switched from the animalistic brain to the decision-making center, choosing to respond instead of react. Good job!

Think of Land as Soon as Conditions Permit as the step to take once you have things under control. This is the long-term solution that brings you back to stability, or your opportunity to choose to change if needed. Jen did this beautifully, even though it was really difficult. In fact, when done right and if conditions permit, it can take a month or so by changing behavior to what serves you well. That's about neuroplasticity.

If your eyes just glazed over, stick with me. Simply put, neuroplasticity is the brain's ability to adapt from an experience by forming new connections between neurons, which are neural cells of the brain and nervous system. This helps to effectively rewire the brain to adapt to new situations, and it's important because it was originally thought that the brain's neural networks stop evolving after childhood.[2] Turns out, though, that brains never stop changing as we adjust to new situations and circumstances—something to remember when imposter syndrome starts to creep in. Through neuroplasticity, we can "change dysfunctional patterns of thinking and behaving" and "develop new mindsets, new memories, new skills, and new abilities," reports *Psychology Today*.³

Dr. Aditi Nerurkar, a Harvard stress expert and the author of *The Five Resets*, says your brain is a muscle, "just like your biceps." You can build it up for less stress and more resilience, in part by strengthening your mind-body connection. She suggests several techniques, including "Stop-Breathe-Be," when you pause before a stressful task to take a deep breath and relax as you let it out. She recommends incorporating it several times throughout your day.⁴

A few months after Jen had her bad review, her boss sat her down and told her how impressed he was that she had taken his criticism and continued to show up. Not only that, but she had proven him wrong.

Ultimately, it had led her to an outcome she wouldn't regret.

Hopefully, you won't have to deal with extreme negative feedback from a boss, but you can apply Jen's method of response to many situations, from parenting decisions to career changes, big purchases to dating.

This is also a good time to apply the Hebbian rule. Proposed by Donald Hebb in his 1949 book, *The Organization of Behavior*, it's based on the idea that "cells that fire together, wire together." So if one neuron helps another to fire, the synaptic weight between them increases, making it easier for it to happen again later.

Picture it this way: A dirt airfield has three runways. Planes land on each of them, but more pilots prefer runway A over B and C. So more often than not, planes land on A, creating a groove that gets deeper and deeper with each landing.

Think of those grooves as brain patterns that affect how we

think and feel. Without them, we'd have to start over each day, like Bill Murray in *Groundhog Day*.

How are your neural pathways being wired? Here are some ways to begin to rewire your brain:

Challenge Your Beliefs: Identify one limiting belief you have about yourself or your abilities. It could be, "I'm not good at public speaking." I used to think this myself and yet, now I regularly speak in front of hundreds—sometimes thousands—of people. Acknowledge it, and then challenge it. Start with small steps to prove yourself wrong. I started by taking a public speaking course that gave me some of the skills I need to craft, rehearse, and deliver a speech.

Learn Something New: Dedicate time each week to learn a new skill or hobby. As you learn, your brain forms new connections, boosting your adaptability. Recently, I went whitewater paddle boarding with some friends in Moab, Utah. I'll be honest, it wasn't pretty, and I've got the GoPro footage to prove it. Just when I thought I'd gotten the hang of it, I'd fall off the board and into the rushing water, and not just once. But I was forging new neural pathways with every paddle—and every dunk in the water. Every little victory boosted my self-confidence. Best of all, I learned something new and I had fun with my friends.

Practice Mindfulness: Even a few minutes of meditation or deep breathing each day can enhance your brain's plasticity. In a study by researchers at the Center for Healthy Minds at the University of Wisconsin–Madison, the brain's circuits exhibited alterations in subjects with thousands of hours of meditation experience. Amazingly, the amygdala showed reduced activity when the subjects were shown emotionally charged images. Mindfulness helps decrease the way emotional stimuli hijacks our brains.

Seek Diverse Perspectives: Engage in conversations with people from different backgrounds and viewpoints. I recommend avoiding the internet for these conversations, where minor disagreements can quickly turn into trolling sessions. Face-to-face interactions with another person, on the other hand, expand your neural networks and broaden your thinking. Engaging in stimulating environments helps improve and maintain neuroplasticity.

Stay Curious: Cultivate a curious mindset. Ask questions, explore new topics, and embrace the unknown. Much like embracing a new activity, such as whitewater paddle boarding, curiosity keeps your brain agile and open to growth. That's because neurons fire whenever you have a thought. When we think about new things, we create neural pathways.

THE MONTH OF TRANSFORMATION

After the bird strike, I struggled for quite a while when it came to sharing the skies with birds. Basically, I would flinch when I saw them. The Thunderbirds have a "hold your ground" policy with birds while we fly in formation. It's dangerous to avoid hitting a bird while you're flying hundreds of miles per hour so closely to other jets. Flying into a bird or birds may or may not have a serious outcome for the pilot, but colliding with another jet always will.

After the bird strike in Colombia, I'd see a flock of birds on a collision course with the formation and I'd use my rudder to move my jet out just a little bit, creating space just in case, and that's a no-no. I could get away with it because I flew on the far left wing with no one else on one side of me. However, the wingmen sandwiched between my jet and the leader of the formation didn't have this luxury.

In a debrief of a flight just a week after my bird strike, it became obvious I was flying wide in the formation. And my boss called me out on it.

"Mace? What's going on there?"

"A giant bird had just passed through our formation," I confessed.

"You need to stop doing that."

I was still in my month of transformation, replaying the what-ifs in my mind. What if I'd ejected? What if I'd lost my engine? I'd had a reality check about the damage a bird strike can do, and I didn't like what I'd discovered. Even with years of training, seeing what could go wrong had messed with my head.

I could have stayed there, stuck in the what-ifs and the oh-craps. Instead, I drew the line, telling myself that I had to fly again the next day, and so I couldn't wallow and let it affect me. The reality check that my boss provided gave me the perspective I needed to move on.

I don't want you to think I just magically changed my perspective on a dime, like Cher smacking Nicolas Cage in *Moonstruck* and shouting, "Snap out of it!" It took me some time not to think about birds while I flew. But I chose to adopt the Big Sky Theory in aviation that says that two randomly flying bodies are highly unlikely to collide. The sky is big. Odds are in our favor. This is likely a naïve theory, but it works well to help muster courage to face an uncomfortable experience.

I certainly had to find the courage to get back in the cockpit and the discipline to continue to execute at the level I needed to fly F-16s for the Air Force. Had I done everything correctly after the bird strike? Maybe not. I might have made a turn to land that was a little sharper than I should have. I could have been more ginger with the jet and taken a more gradual bank, not knowing the extent of the damage.

These were all things to consider for how I'd respond if I ever had a bird strike again. But the month of transformation was less about the actions I had taken in the air and more about getting back on the horse, so to speak—a horse with some twenty-nine thousand pounds of thrust. It was about my mindset as transformation so often is.

FLIP YOUR MINDSET

The first few times I flew in a Thunderbird jet, it felt familiar and comfortable and, at the same time, not. I'd flown a thousand or more hours in the F-16, and I knew how it responded to my piloting. But the Thunderbird skills are so different, because now you're flying in close formation and often upside down. So at first, I kind of sucked at it. We all did, and that's disconcerting, because you have to be an accomplished fighter pilot to fly for the Thunderbirds. Yet we looked like we were learning—because we were.

Whenever I struggled with a maneuver, I doubted my capabilities, but my instructor would assure me, "Yeah, that's totally normal." We regularly referred to a new pilot's first training season as a game of Whac-A-Mole. You'd feel like you were getting the hang of a new maneuver, but the next day you'd go out and struggle with it again. This could be frustrating and demoralizing even though I knew I was learning difficult skills that took a lot of practice to master. Of course I struggled! But I kept at it, and by my second year, I took on the role of the instructor teaching newbie Thunderbirds the complicated maneuvers we were known for.

This takes a *growth* mindset, when we believe that our abilities and our intelligence can be developed over time. It's the opposite of a *fixed* mindset, which is the assumption that these things are unchangeable, that we're born with a static set of talents and skills. A fixed mindset says, "Why bother?" and a growth mindset asks, "Why not?" Both mindsets are ultimately based on how comfortable you are with change.

This idea was coined by psychologist Carol Dweck in her book *Mindset: The New Psychology of Success*. Dweck wrote that people with fixed mindsets avoid challenges and obstacles, ignore criticism even if it's useful feedback, avoid effort, and envy others' success. People with growth mindsets, on the other hand, embrace challenges, persist despite obstacles, learn from criticisms, believe effort is the way to reach mastery, and learn from others' successes.[5] When it came to my call sign, I could have chosen a fixed mindset and failed to grow from my dangerous mistake. It took a while, but I made the choice to learn from my error and, ultimately, I grew from it.

In one of Dweck's studies, her team analyzed the brain activity of students as they reviewed mistakes they'd made on a test. The fixed mindset students showed no brain activity, while the growth mindset demonstrated processing activity. In short, they were learning from their mistakes.

People with a fixed mindset tend to respond to failure by withdrawing. Quitting. People with a growth mindset see failure as a lesson to learn so that they can improve next time. Just ask the brilliant minds at the Massachusetts Institute of Technology.

THE FLIPSIDE

MIT's Teaching + Learning Lab reported that fixed and growth mindsets had completely different attributes in several areas:

Effort:

- Fixed mindsets see effort as a sign of weakness.
- Growth mindsets see effort as important for learning.

Failure:

- After failure, fixed mindsets either determine they lack ability or blame others or the situation.
- After failure, growth mindsets objectively identify their role in it, such as lack of effort or preparation.

Feedback:

- Fixed mindsets avoid feedback and act defensively.
- Growth mindsets seek out feedback and use it to make improvements.[6]

My decision to leave the Air Force was rooted in a growth mindset. It wasn't that I was leaving the military to get away from something, although my aching back from all those G-forces might disagree. I liked being a fighter pilot. It was that I was pulled toward something new and exciting. It felt the same as seeing those jets taking off at Tyndall Air Force Base and deciding I wanted to fly F-16s. It was a new challenge, and I knew it would require me to, once again, become a beginner.

It took a while for me to get to that place, because having to work really hard can make you feel like you don't belong, and that was exactly how I'd felt in F-16 training. I thought everyone knew more than I did, so I studied hard, but it just wasn't enough on its own. I also felt very reluctant to ask questions because I didn't want to look clueless. But that's counterproductive to the growth mindset, and it sabotaged my progress and my confidence.

To me, a growth mindset is about a devotion to ever-evolving goals wrapped in a willingness to take chances, rather than taking the path of least resistance. This is where you have to take ownership and refuse to fall into the victim mentality where you blame other people or the situation for your failures. Even if the culture is working against your growth, your boss doesn't manage well, your marriage is stagnant, or the situation isn't ideal, you need to focus on the things you can control; this is the secret to enduring so you can grow. (See chapter six.) Even though the growing pains were, at times, excruciating for me, they were not as painful as "comfortable misery" became.

How I talked to myself affected my mindset, too, and as my career advanced, I got better at this. What you tell yourself becomes your belief system. It's like what Dr. Marie Claire Bourque, who works with athletes and others on their performance mindsets, says: "Language matters." She explained that your belief system in turn informs your reality, so she wants you to say kind things to yourself. But positive self-talk is more than an inner cheerleader, it's also a reframing of all the incorrect and unhelpful things that your inner critic says. Helpful self-talk takes a negative thought like *You flew faster than you should have. You're a lousy fighter pilot*, and reframes it as *You're new and you got overwhelmed. This is what training is for.* When I was in an emergency situation in the cockpit, taking a breath and telling myself *You know what do. You've trained for this* allowed me to prevent panicking so I could focus on the appropriate response.

To me, that's more of a reality assessment than positive self-talk, but they do go hand in hand. It's not a bunch of rainbows-and-unicorns mantras. These statements are based in reality, like a narrator calmly explaining what's really happening.

Instead of *You suck and you'll never amount to anything*, it's *You're human and you made a mistake. You'll learn from it and do better.*

Or in lieu of *You're not smart enough*, say to yourself, *You don't know enough yet, but you will.*

"Yet" is such a powerful word. It's a way to give yourself grace while you earn self-mastery. If you watch videos of famous athletes, actors, or musicians when they were first starting out, you'll discover why "yet" matters so much. Google "before they were famous" and it's likely you'll find a video of a preteen Miley Cyrus auditioning for *Hannah Montana*, Marshall Mathers's first recorded performance on faded videotape in what appears to be a sparsely attended outdoor concert two years before he was signed, and a baby-faced Stephen Curry shooting hoops with his little brother. They weren't "yet" the Miley, Eminem, or Steph we now know. But someday, they would be.

When I started training with the Thunderbirds, I didn't know how to fly upside down, put my body through nine times the force of gravity just one hundred feet above the ground, or act as a role model to the millions who would see us throughout three seasons...yet. I had to learn how to do all that, and throughout it all, I had to talk myself through it.

When life hits you hard, act like you've got to land a plane that has holes in it. Respond instead of reacting by taking the steps for enduring an emergency: (1) Maintain "aircraft" control. (2) Analyze the situation. (3) Take proper action. (4) Land as soon as conditions permit. That will keep you out of your animalistic brain and in the part of the brain that gets things done. Along the way, give yourself the grace you'll need to transform and, please, be kind to yourself. You're still learning. We all are.

Scriptside:

- Letting the animalistic brain stay flooded, keeping your brain in the six *F*s.
- Getting stuck in analysis paralysis.
- Taking too much time to respond, analyze, and transform.
- Choosing a fixed mindset over a growth mindset.
- Being mean when you talk to yourself.

Flipside:

- Prioritizing in an emergency.
- Giving yourself grace and moving forward.
- Taking one minute to respond, one hour to analyze and reflect, one month to transform when possible.
- Using a growth mindset to transform and respond better next time.
- Talking to yourself with kindness.

CHAPTER SIX

Don't Get Stuck in the Comfortable Misery

I probably shouldn't have married my high school sweetheart, but I did. After a long-distance relationship through college, we fell into the trap of expectations for what you're supposed to do next: We got married. We were from the same small town, had the same friends, and our families knew each other. We followed the path of the comfortable.

Suddenly, he was an Air Force spouse stationed in Japan where the only social network he had was the other military spouses—all women. When you're a military spouse, you are essential to the morale of the combat unit, but there's a lot of pressure for your whole identity to center around that role. You move every three years, making it difficult to put down roots, build your own friend groups outside of the base, or grow a career. The military even calls you a "dependent," and that's hard for anyone, but especially for men who might feel like they're supposed to be the provider in a relationship.

Meanwhile, I was a fresh-faced fighter pilot struggling to learn my new job and prove myself. My husband and I never fought and we weren't really unhappy, but we were basically on autopilot. We showed up to squadron events looking like a happy couple, but my job had become my sole focus. I was spending twelve to fifteen hours per day at the squadron, and then at home I was still mentally at work. He saw me growing as a person, constantly forced outside of my comfort zone. Meanwhile, I worried that he felt stagnant, emasculated by my career and what appeared to me to be his own lack of purpose. Soon it became very apparent that his confidence levels dropped, in part with help from me. I found myself wondering, *This can't be it, can it?*

We'd dated in high school, stayed a couple while apart for college, and then we got married. So what was the next logical step? Right, having kids. It's a big decision for anyone, but for a female fighter pilot, it comes with a huge career shift, because, at that time, once you were pregnant, you were pretty much out of the aircraft for a year. I had a lot to think about, so I was waiting, waiting, waiting, and suddenly, I had a realization: I didn't want to have children with him.

One afternoon, in the middle of an otherwise unremarkable conversation, I remember thinking that he and I were in two completely different places in life, and that I wanted my children to be where I was, not where he was. *My* children. I couldn't even picture us having kids *together*, not because there was anything wrong with him. Rather, it was what was wrong with us, as a couple. We'd come from the same place, but we weren't headed in the same direction.

I wanted different things than he did. I was always chasing the next achievement. I wanted to prove that I deserved to be in the fighter pilot community. I needed to be better at my job, to be stronger and faster. Meanwhile, I noticed little things he'd say in conversation or how he'd talk to his family on the phone, and it became clear that what he really wanted was a house and kids back home in Wisconsin, and I knew that would never be enough for me.

For four years, I had been in an intense training program designed for accelerated personal growth. I mean, if the Air Force is going to trust you with a $20 million jet, they're going to want to make sure you get the mindset for it and get it fast. I'd evolved from fighter fan girl on the tarmac in Florida to real live fighter pilot stationed in Japan. Along the way, I competed for and got that coveted fighter jet. Then, I had a "that could have killed me" experience that earned me my call sign, all the while setting aside my softer side to survive in the testosterone-fueled environment I was in.

My husband had about as good a job as a military spouse can get on base, as a branch manager for the credit union, but he was restricted in his work options. So while my high-performance career took off like my jet, requiring me to force myself out of my comfort zone day in and day out, his stayed predictably steady. He'd met me when I was Michelle, but now I was Mace, and he wasn't coming with me on the rest of this journey.

This divide began to weigh on me. I would think about it whenever we were together, and soon, it became an all-consuming burden. It was like a switch had been flipped on and now there was no going back, like I'd taken the red pill in *The Matrix* and couldn't unsee what I'd seen or unlearn what I'd discovered. We didn't have a bad marriage, and I didn't dislike him, but I knew that we didn't have a future together. It would have been easier if he'd been a terrible spouse, but instead, I was about to hurt someone I cared about.

I decided to make a move that required the most courage of my life—I asked for a divorce. People were shocked, especially our families. We'd both grown up in the same small town of a few thousand people where everyone knew everyone, and everyone knew our business.

Worse, he didn't see it coming, and in that moment when I asked for a divorce, I took all the fear and stress off my shoulders and placed them on his. At first, I felt relieved for unloading this secret I'd been carrying but it soon turned into guilt... and shame. To me, this was a failure, a very public failure. I had stumbled before, but for the first time, I was falling on my face. I believed that a failed marriage wasn't something that happened to people like me—my parents' easy, golden child, the high-achieving, straight A student—even if I was the one who chose to end it.

I struggled with conflicting feelings. On the one hand, I was relieved after asking for what I knew I really wanted. Yet, I also felt grief over losing our relationship and the person who was so much a part of my young adulthood. All of this was wrapped in guilt for hurting someone I cared about and shame for being a failure.

We had been part of a social group made up of other married couples on base, and suddenly, I didn't quite fit into that group anymore; I hadn't established friendships with any single pilots. I was going through a traumatic time in a fishbowl. There was no separation between my work life and the rest of my life, which felt like it was in shambles for all to see.

One afternoon, we were both at home during a workday, slogging through the logistics of the divorce, and I wondered who could see that our cars were parked outside. Everyone, that's who, because we all lived in the same neighborhood. Suddenly, it felt

THE FLIPSIDE

like we were starring in one of those *Housewives* reality shows, and there was no hiding what was going on.

He soon moved in with friends of ours and got ready to leave Japan, while I found solace in working out, drinking wine, and taking hot baths. That doesn't sound so bad until you're drinking a bottle of wine by yourself and sitting in the bath staring at the wall until the water has gone cold. On a Tuesday.

After he left for the States, I began my life alone for the first time ever. On Christmas, I went to Smokin's place, and her five-year-old daughter wanted to test out the makeup kit Santa had left under the tree on "Miss Michelle." So here I was, my first Christmas alone, mulled wine in hand at 10:00 a.m., letting a kindergartener paint my face. And I mean, *paint*: black mascara on my cheeks and red lipstick on my forehead. I. Did. Not. Care. I thought, *This is my life now*. All I could do was laugh at the ridiculousness of the moment.

But on other days, the situation got very serious. I'd cry in the shower, I'd cry alone in my apartment, and I'd cry on my drive to work. One morning, I was driving my little Daihatsu Opti classic, a tiny car like a Volkswagen Beetle, on a winding hilly road overlooking a cliff, and hopelessness washed over me. I felt like I was failing and—my biggest fear—everyone was watching me do it. Life on base meant that there was no separation from work or family and therefore no separation from my personal problems. I didn't even interact with supermarket cashiers or waiters off base because I didn't speak the language. I felt trapped in this foreign country and I felt completely alone.

As I neared the edge of a cliff with a plunging drop-off toward the lake below, I had a thought. A terrible, intrusive, desperate thought: For a brief moment, I considered driving off the cliff. It seemed like the solution that would end my pain. Then I thought through what that would be like, to actually drive off that cliff, what that would really mean. I pictured the Air Force knocking on my parents' door, and I stopped myself, keeping my hands steady on the wheel.

In hindsight, I realize I should have sought help, but I didn't have the courage to tell anyone how much I was struggling, and back then, the military didn't embrace mental health support. I was experiencing situational depression, a stress-related type of depression brought on by a specific incident, and I was struggling. But at work, my commanding officer expected my divorce to go on in the background while I was on my game for thirteen-hour workdays as a "contributing member" of the squadron.

Honestly, it would have been easier just to stay married. We'd been together since junior year of high school, since we were *children*. There were times during college hundreds of miles apart that we could have easily decided, "Okay, this relationship has run its course." But we didn't. We'd put so much time into the relationship, I don't take marriage lightly, and it wasn't a bad relationship. It was just *fine*. And it's easy to choose fine, to pick comfortable misery, no matter the stakes because we're often likely to resist change.

WHY WE RESIST CHANGE

Imagine you have a heart condition and your doctor tells you that if you don't change your habits—quit smoking, lose weight, exercise more—you'll die. Dramatic, I know, and the solution seems easy: Make changes right away. But Harvard researcher Lisa Lahey reports findings that only one in seven at-risk heart patients could successfully follow through.[1] That's just about 14 percent reportedly choosing change over death!

It's tempting to decide that the six in seven in the study were just plain lazy, but Lahey, who's a coauthor of *Immunity to Change*, said it's about more than deciding to "just do it," as Nike would say. "Change or die" is what she calls the New Year's resolution model of change. It involves willpower, and sometimes that works. But for those who don't change, there's usually something more complex happening behind the scenes.

Lahey says that our built-in resistance to change is our brain's way of protecting us from anxiety and fear. We don't like uncertainty and we don't like to lose control, so instead, we choose comfortable misery, which could be called the Devil You Know Syndrome. Change is hard and so is staying stagnant, but at least you're familiar with one of them.

We humans simply don't like the unknown. A study by University College London found that people who were told that they had a 50 percent chance of receiving a shock were more stressed than people who were informed that they had a 100 percent chance of being shocked.[2] One of the researchers said, "Our experiment allows us to draw conclusions about the effect of uncertainty on stress. It turns out that it's much worse not knowing you are going to get a shock than knowing you definitely will or won't."[3]

If I'd stayed married but unhappy, I would have known what I was in for. But getting a divorce? That was full of uncertainty about how my husband would react and then respond, what people would think, what it would be like to go to my hometown afterward, and what dating someone new for the first time since high school would be like.

Though I did the hard work of choosing change by asking for a divorce, that's when I got stuck. Consumed by a feeling of failure, I found myself spiraling with no perspective and no one to ask for help. Of course, I expected to go through a rough patch coming out of a significant and stressful life event like a divorce. But there was a point where I should have chosen to take ownership of my decision and the role I'd played in a failed marriage and set an intentional path forward.

Instead, I felt sorry for myself. I withdrew from the people around me and gave up the things I had control of: my actions, my attitude, and my response. I ended up going along for the ride on a tsunami I had created. All I could do was hang on and hope I eventually emerged on the other side in one piece.

What kept me stuck in extended wallowing was shame. I was embarrassed by my divorce because it felt like a personal failure. Brené Brown, who's done extensive research on shame, says the emotion has two main messages for us: "You're not good enough" and "Who do you think you are?"

There are several flavors of shame, including social, moral, and personal, and mine was a triple scoop. I was ashamed of how others were perceiving my divorce, ashamed of divorcing when I'd taken vows till death do us part, and ashamed that I was the one who initiated it. Then, after all that, I got stuck. Truthfully, the divorce was way harder than I'd anticipated. The hardest part wasn't the loss of the relationship or being alone. It was the feeling of failure that came with it. When I called my parents to give them the news, my mom started crying. Then she asked, "Did we do something wrong when we raised you?"

Oof. Yet in the long run, the divorce brought my parents and me closer, and I confided in them a lot in the months to come. I guess we all had our own feelings of shame to work through, and it was better to work through them together. I mean, they had a decent shot at running into my ex or his family in the supermarket back home, whereas I had half a world between us.

Time heals, and I was certainly taking a lot of it, much more than the "one month to transform" that I now promote. There's a reason for that. The National Institute for the Clinical Application of Behavioral Medicine says that shame creates a fear of rejection that can keep people trapped in any number of avoidance strategies used to escape the pain.[4] Then the shame grows even more powerful.

Shame kept me from seeking help and from sharing my feelings with most everyone I knew. It felt like airing my dirty laundry, and why would I do that when I was trying to impress the other pilots in my squadron and my commander? So I kept my mouth shut, and they walked on eggshells around me.

This may sound silly, but my biggest supporter went back to the States with my ex: our golden retriever, Charlie. He couldn't stay on base with me because my hours were too long, so he was sent home right when I needed him the most. You know how retrievers sense you're sad, so they place their head on your lap and raise their eyebrows and stare into your eyes empathically? God, I needed that.

Plus, Charlie was a big part of our lives. A bunch of friends from my squadron also had goldens, and we'd all meet on the golf course at the end of our street for dog playdates several times a week. But with Charlie gone, that connection, too, was gone.

I saw Charlie again just once, years later, in the background of a photo. I winced when I stumbled across him online, and for a few seconds, I avoided looking at him before closing the webpage. I didn't want to risk ripping off an old emotional scab when I wasn't ready.

As I stumbled through the divorce, I needed an outlet, so I put my energy and fear and shame into exercise. Extreme exercise. I'd sprint so hard that I'd end up dry-heaving into a ditch on the side of the road all too often. Then I'd forget to eat from the stress, so I dropped a lot of weight quickly. I weighed less than I had in high school, and it wasn't healthy.

I'd been through stressful training programs, but this wasn't like that at all. At least I shared that stress with other trainees and had something to be proud of waiting at the end. Here, I was alone, and my whole world felt like it was falling apart. I hadn't yet realized that I could apply everything I'd learned in the aircraft to my life. I didn't give myself a deadline to stop wallowing and start transforming, and I couldn't overcome my inner critic, who liked to remind me every day that I'd failed and then threw me pity parties. I didn't even begin to pursue the things I'd envisioned I'd be able to achieve after the divorce.

It took over a year and a move to an Air Force base in Texas for me to begin to grow again—and grow fast. That was when I discovered that bad things can catapult you to great things if you apply what you learned and choose how you respond.

POSTTRAUMATIC GROWTH

We've heard a lot about posttraumatic stress disorder, especially when it comes to the military, but what about posttraumatic growth, or PTG? Richard Tedeschi, PhD, among the founders of PTG research, defines it as "positive psychological changes experienced as a result of the struggle with trauma or highly challenging situations."[5] In other words, you go through something bad and then evolve *because* of it, not *in spite of* it.

Like Nietzsche and not a few internet memes have said, "That which does not kill me makes me stronger." The data support it, at least for certain people. Tedeschi and his colleagues reported several ingredients that make up growth after a traumatic incident, including:

- Seeing yourself as a survivor instead of a victim.
- Self-reliance, a sort of "If I survived that, I can survive anything" feeling.
- Heightened awareness of vulnerability and mortality.[6]

They identified five key PTG areas: personal strength (resilience, confidence, authenticity), spiritual growth (deeper awareness and faith), new possibilities (interests and perspectives), improved relationships (belonging, vulnerability, empathy), and an appreciation for life (gratitude, clear priorities).[7]

When you achieve PTG, you turn a bad experience into a learning experience and come out stronger for it. David Sanderson, one of the passengers on the "Miracle on the Hudson" plane that was taken down by a bird strike, refers to incidents that force you to rethink your life as "jolts." He says that they can either derail or propel you, depending on how you interpret the event, and he likens them to "personal plane crashes."[8] As much as I want to resist the obvious pun, I'd had a pretty big "personal plane crash" myself, and it didn't involve my F-16. Not directly, anyhow.

My time in Texas helped me get past it and toward something closer to PTG. It catapulted me into a feeling of new possibilities.

I had been very busy playing dress-up by pretending to be happy in a relationship where I really wasn't and pretending to be a confident fighter pilot when I was the opposite—so busy that I wasn't being my authentic self, and that made me question my identity, which shook up my resilience. But in my new squadron on a new base, it was immediately apparent to me that this unit accepted people with different likes and dislikes. They looked homogenous, but they had diverse hobbies and priorities outside of work. I finally felt like I could show up authentically.

I was still carrying the tail end of the shame and embarrassment around my failed marriage, and I felt like that had become my identity in my squadron in Japan. In Texas, though, no one knew about any of that. Most didn't even know I'd even been married before. They didn't know the details of how I'd earned my call sign, or about the weeks and/or months when I'd shown up at work to just go through the motions while being consumed by my own emotional turmoil. It was like a do-over, and I was happy to take it.

For the first time in months, I had faith that everything was going to work out. After all I'd been through, I felt like I could handle anything that was thrown at me, and I knew I'd figure it out. The name of Glennon Doyle's podcast sums it up nicely: *We Can Do Hard Things*. These hard things were much harder than the little and not-so-little setbacks I'd endured in training: a near G-LOC, a mental fight through claustrophobia, uncomfortably being singled out again and again. All of that was happening while the most important relationship in my life was eroding. The smaller hard things layered with my marriage breaking up, and it felt like a dome that covered every aspect of me. Yet I'd proven my own strength to myself.

THE FLIPSIDE

I didn't know it back then, but I was checking off the PTG list as I went along. Personal strength? Check. Spiritual growth with its awareness and faith? Check. Improved relationships? My parents and I were never closer before the months following my divorce. And, in Texas, I finally felt like I belonged.

So much of my shame felt tethered to the locations and people on the Air Force base in Japan. Moving frequently in the military is often tough, but this time it was a chance to rewrite my story in a new squadron and a new city. Except for one thing: My new squadron had never had a woman fighter pilot as part of it before. Ever.

I didn't want this fact to change how I felt as I showed up at my new squadron, but I'd be lying if I said it didn't. The first time I'd meet them all was at a roll call. Was this where I wanted to meet my squadron after my prior experiences with roll calls? No. Nervous, I steeled myself to deal with the uncomfortable comments or stonewalling I thought was inevitable.

Boy, did I get a surprise: Everyone was welcoming. They didn't just tolerate me as part of the unit, they immediately made me feel like I belonged. When my name was announced in roll call, everyone cheered. Throughout the evening, the other pilots came up one after another and introduced themselves and asked where I was coming from and whether I'd found a place to live. Plus, there were no bouncing boobies anywhere.

Empowered by this new support network and new lease on life, I began chasing dreams that had previously felt unreachable. Six months after I moved to Texas, I went to Kathmandu with total strangers to walk eighty miles through the mountains. I'd had a long list of hobbies I'd always wanted to pursue but I never went after them because in Japan, I felt like I was always at work, even when I was at home. Plus, these hobbies were either inaccessible while living overseas or they were things my husband hadn't been interested in.

In Texas, many of the pilots flew F-16s part-time and flew for the airlines as their primary job. That meant that the squadron couldn't rely on consuming every part of people's lives to be tactically effective. We were good at what we did, but everyone from the top down understood that could be true while still giving people time with their families and time to pursue interests outside of work.

So, I picked a place I'd always wanted to go, requested leave as I'd saved up weeks of time off, and bought a plane ticket to go do something I'd wanted to do since watching adventure documentaries with my father: trekking to Everest Base Camp.

Everest is a lengthy and prohibitively expensive endeavor, and I couldn't take months off the Air Force to train. But I'd always wanted to go to Nepal and see the mountain. Besides, I'd traveled from the United States to Japan alone several times, navigated the Tokyo train station even though I didn't understand the language, and I'd been living solo for a while by then. So why not?

Nepal felt like a place of adventure, possibility, and serenity. I had countless breathtaking moments where I had to pause to literally catch my breath, not just because of the elevation—at times, we hiked above sixteen thousand feet—but because I realized how lucky I was to even be in such a beautiful place.

One of our stops along the way was the tiny village of Tengboche. Surrounded by the towering snow-capped peaks of the Himalayan mountains, the village isn't much more than a tea hut and a monastery, where we were invited to join the monks for a ceremony.

There, our group of trekkers filed into the prayer room and took seats along the wall. The monks entered, chanting and banging a gong. Maybe it was the altitude, the exhaustion, or the sudden warmth after hiking miles in the cold, but what I felt sitting in that monastery was the closest thing I can describe to a moment of enlightenment. And I've never even been especially religious.

My whole body began to tingle, and it was like I realized that the chapter of struggle I'd been slogging through was now over. Suddenly, I saw all the possibilities that existed for me in the world. At the risk of sounding like an excerpt from *Eat, Pray, Love*, that trip gave me the closure and the perspective I'd so desperately needed.

When I got back to Texas, my squadron had a lot of questions about my adventure. They wanted to see the videos and the photos, and they asked for detailed stories. It seemed they were all living vicariously through me, because most of them had families they couldn't leave for three weeks to traipse around Nepal. I became the girl in the squadron who did adventurous, badass things. They'd even introduce me to their daughters as "This is Mace. She flies jets with Daddy and climbs mountains, too."

I'd become me again. Me, only better.

Your "mountain" doesn't have to be a literal mountain, and you don't have to go on a grand adventure, chanting with monks, to achieve PTG. Even a small change can help you begin to turn a bad experience into a good outcome that ultimately leads you to growth. For instance, when you feel shame rearing its ugly head, take a moment to acknowledge it instead of trying to stuff it down deep inside. Or when you're triggered by something someone said, catch your breath and choose a calmer response Think of ways you can escape your comfort zone, big or small. Maybe you can go for a walk instead of doom-scrolling on the couch, take on a new hobby, or plan your annual vacation somewhere new to you instead of taking the same old trip.

THE STRESS TEST

Divorce, even if you're the one who initiates it, is a major life stressor. It ranks number two on the Life Change Index Scale (aka the Stress Test) after the number one stressor, death of a spouse.[9] The scale is a checklist, created in 1967, of life's most stressful events, including serving jail time, getting fired, illness or injury, and death of a close family member.

If you've endured big stressors, you're at risk for depression. In fact, people who develop depression are between 2.5 and 9.4 times as likely to have experienced a major stressful life event,[10] which is really no surprise. UCLA School of Medicine professor of psychiatry Daniel Siegel, MD, says, "Relationships are the most important part of our having well-being in being human. It's that simple. And it's that important."[11] If the most important relationship in your life dissolves, it will take its toll, as it did for me.

But relationships aren't the only cause of stress, and most of us will endure a major stressor at one point or another. *Scientific American* states that about 61 percent of men and 51 percent of women in the United States report at least one traumatic event in their lifetime, though that study was conducted long before we all went through the pandemic together.[12]

After something bad enters our lives, some of us seek meaning, and in doing so we can come to recognize the growth that came from the trauma. It's even happened to military veterans who had

PTSD. The National Health and Resilience in Veterans Study found that nearly three-quarters of veterans who screened positive for PTSD reported at least moderate PTG from their worst traumatic event.[13]

The thing about PTG is that it kind of just...happens. But we can speed it up. Tedeschi wrote in *Harvard Business Review* that there are five elements of such growth:

1. **Education:** Learn about the trauma that rocked your world.
2. **Emotional regulation:** Manage negative emotions.
3. **Disclosure:** Talk about it with someone you trust.
4. **Narrative development:** Determine how the trauma made you stronger, a better person, more appreciative, etc.
5. **Service:** Find a way to help other people going through the same thing.[14]

Turns out that plenty of people, veterans and civilians alike, are psychologically resilient. We bounce back eventually. But posttraumatic growth and resilience aren't the same thing. I've come to think of resilience as a part of your personality, a personal characteristic independent of any specific event. I'm not the only one who thinks that way.

Kanako Taku, PhD, survived the 1995 Kobe earthquake in Japan and also researched PTG. An associate professor of psychology at Oakland University, Taku told the American Psychological Association, "Resiliency is the personal attribute or ability to bounce back." PTG, on the other hand, involves a long process that "takes a lot of time, energy, and struggle."[15]

So while PTG is a good thing, resilience is what you need to be able to transform as quickly as I've proposed. My divorce gave me PTG, but resilience was something I had in me all along. I just needed to figure out how to use it.

Scriptside:

- Choosing comfortable misery over change.
- Stunting your own growth to stay small because it's less painful.
- Avoiding the lessons that come from hard times.

Flipside:

- Making the hard decision to choose growth.
- Allowing posttraumatic growth to teach you better ways to be.
- Doing the hard work of change.

CHAPTER SEVEN
Seek Silver Linings

After I secured my oxygen mask for the Air Force's high-altitude-chamber training, I wondered if my fellow airmen—and they were all men—could tell by my eyes that I was panicking. I was merely a few weeks into pilot training, and I was running into trouble. The training teaches future pilots how to recognize the early signs of hypoxia, when the brain lacks enough oxygen to perform basic functions, like identifying shapes and colors. Everybody experiences symptoms of hypoxia differently, and this training allows you to find out what your first symptoms would be by depriving your body of oxygen. That way you can take steps to prevent its progression before you get to the point of, say, thinking a circle is a square and that red is purple. If hypoxia is allowed to go unchecked, we can become completely disoriented while flying a jet at the speed of sound.

But I had only gotten as far as pulling on the oxygen mask, the one every fighter pilot must wear in the cockpit. At that moment, I had a choice: Relieve my claustrophobia by pulling off my mask or push through the panic. Each decision came with a price. If

I removed my mask, I'd abandon months of training and take becoming a pilot off the table. I'd be rolling the dice with my future while the Air Force decided what to retrain me for, such as intelligence, maybe, or maintenance. But if I kept it on and endured my panic attack, I'd be one step closer to my dream of flying jets.

Pulling off that mask would have been a huge failure, because it meant that I couldn't overcome my own fear in a situation that wasn't actually dangerous. I wasn't suffocating in the mask. I was creating fear where there didn't need to be any, and if I couldn't power through that, I surely couldn't fly an F-16.

I knew that the price of quitting was too high. I already felt the pressure of needing to prove that I could survive, and hopefully thrive, in a career that regularly raised eyebrows. So I talked myself out of my hyperventilation and calmed my heart rate. I'd kept my eye on the reason I was there and on the silver lining—the good thing that comes from something difficult or unpleasant—and passed the test.

You can think of the silver lining as pretty much the light at the end of the tunnel, the puppies and rainbows, and the Willy Wonka factory golden ticket wrapped in one. If you can keep aiming for it, you'll be able to endure so much now that will allow you to enjoy so much more later. But it's more than the reward at the end. It's also what you can take with you from the experience itself.

RESILIENCE MEETS REGRET

In that moment when I wanted to yank off my oxygen mask during high-altitude training, resilience and regret faced off briefly. I had to be resilient or deal with the regret. If posttraumatic growth is something that happens to you when you look back at tough times, resilience is something you *choose* while you're in it. It's a muscle that you build. When it comes down to it, resilience is mental toughness.

The military knows a thing or two about building resilience. Whether it's basic training or the much more intense Basic Underwater Demolition/SEAL training, where elite Navy trainees skirt hypothermia in extreme conditions, it involves both physical and mental endurance. But it isn't just extreme situations that build resilience. Sometimes it's the everyday slog.

When I showed up in Misawa, Japan, it was winter, and winter flying in the tiny cockpit of an F-16 is uncomfortable. We had to wear many layers because we were flying over the near-freezing waters of the Pacific Ocean, and we had to be prepared for the remote possibility of needing to eject and land where those layers would be required to keep us alive. The first layer of clothing after underwear is a cat suit, which is like a fleece onesie you might wear under your ski gear.

Next, the infamous poopy suit, a waterproof anti-exposure suit with a rubber seal on the wrists and around your neck that feels like it's choking you. We had to burp the suit—squeezing the air out as if our bodies were leftovers in a Ziploc bag—but it was never enough and we'd wind up with air in the attached feet (think, footie pajamas), which made putting on your boots damn near impossible. So we kept a Shop-Vac in our locker room, and we'd stick the hose under the suit and suck out the excess air, effectively vacuum-sealing ourselves in. Next was the flight suit and then a survival vest, which had a radio, flares, and other essential gear in it. Then you'd put on the G-suit, which had the air bladders that squeezed your legs to keep your blood flowing. Finally, all of that would be topped off with the harness that attached to the aircraft, and then your gloves and your helmet.

We all looked like the Michelin Man. I thought, *Is this normal? I can barely move.* But it was. This was how all the other fighter pilots suited up every day, and it's how I would, too. By the time we waddled to our jets and strapped in, we'd be sweaty and out of breath.

Then we'd have to do our ground ops, which involved checking several switches and running important systems through preflight tests. I'd have to check a few knobs and switches by my left hip, and whenever I turned my head, the rubber around my neck would cut off my breathing. So I'd have to take a deep breath, pull my mask to the side, and quickly try to see all the things I needed to check. Then I'd catch my breath again before moving on to the next step. And that was before we even took off!

That cumbersome, multilayer ensemble practically choked me, in the tiny F-16 cockpit, and gave me feelings of claustrophobia, as though I was trapped, overheating, and I couldn't quite catch my breath. I wondered, *How do you guys do this?* No one else seemed to be talking about how suffocating the layers of equipment felt. During many early flights, it would feel like I wasn't quite getting a full breath of air while sitting in the cockpit waiting to take off, and I'd consider if something was wrong with my mask or the jet's oxygen system. Yet after a once-over, everything always seemed in order. I realized it was just me—the problem was me—and the more I focused on the alarming feeling of not being able to breathe, the worse it seemed to get.

But you know what? I got used to it. It took several flights and months of wearing the full winter attire before my thoughts shifted from *I can't catch my breath* to *It's not that bad*. At first, changing my mindset took a conscious effort but eventually, I thought about it less and less. I chose not to let myself freak out and let the uncomfortable outweigh the reward, just like that day in high-altitude training when I chose to keep the oxygen mask on.

Resilience is a choice. You have to do hard things to become tough, and you don't have to be in the military to need or to create mental toughness. There are plenty of situations in civilian life where it's necessary and where it can be cultivated.

BOUNCING BACK

I've experienced posttraumatic growth but I've also learned a lot about my own resilience, the process of adapting to that trauma. The stuff you have to dig deep to find and use to get through the hard times.

I say "dig deep" because I believe that the capacity for resilience is a part of who you are, and it lives inside you, ready to be deployed as needed. It might not be something we're born with, but it can be cultivated, nurtured, and expanded.

The American Psychological Association says that resilience is a process of adapting to challenging life experiences and cites three factors:

1. Our view of and engagement with the world.
2. Our social resources.
3. Our coping strategies.[1]

I suppose that it's things like whether we see the glass as half-full or half-empty, how much help we have from the people in our lives, and what we use to cope. Put those all together, and, I believe, you have the ability to find the silver lining even in the bleakest of situations.

Back in the 1950s, a Harvard researcher named Dr. Curt Richter conducted a (rather cruel) study on rats, both domesticated and wild, to measure how long they would swim in jars before giving up. The wild rats all died within minutes of entering the water, but nine of the dozen domesticated rats swam for days before giving up.

So he adjusted the experiment. This time, he waited until just before the rats were about to die, and then he plucked them out of the water and held them gently for a while before putting them back in. After that, the wild rats would swim just as long as the domesticated rats, evidently finding grasping onto the good thing that happened despite the bad. When they were returned to the water, they had hope that a good thing might happen again. This helped them build resilience to endure the experiment much longer.

Of course, we have far more control over our lives with situations typically far less dire, but the rat's endurance proves how much weight silver linings can carry. Even in your darkest hours, digging deep will help you push forward regardless of how hopeless you might feel.

CAN FORESIGHT BE 20/20, TOO?

If I were to ask you to think back about the good things that have come out of the bad things that have happened to you, I know you'll be able to list something. But that's hindsight. What about foresight?

I believe that foresight comes down to being realistically optimistic. I say "realistically" because too often we get trapped in the power of positive thinking that's all over social media, and then we end up in denial, a sort of la-la land where you try to talk yourself into becoming a tech billionaire or the next Taylor Swift if you "just imagine it." Psych Central says, "Realistic optimists are cautiously hopeful of favorable outcomes, but they do as much as they can to obtain the desired results."[2] It's more than a positive mindset. It's the work behind it, too. Though it's critical to have a vision of where you want to go, if you stop there without coming up with a plan, putting in the work, and acknowledging any obstacles, the odds are not in your favor. Take my friend Grant Korgan, for instance. He's the most positive person I've ever met, and when I hang out with him, my whole perspective on life shifts for the better. I walk away feeling energized because Grant is like a battery charger for humans.

But Grant has had to work hard on his optimism. In 2010, he was snowmobiling through the Sierra backcountry when he took a jump and overshot his landing, fracturing his L1 vertebrae, a rare injury to the spinal cord that left him paralyzed from the waist down. He'd gotten married just a month earlier, and now, he was a paraplegic.

But today, Grant surfs... on a specially modified board, under the barrel of a wave. He wake foils, seated on a small board that's pulled behind a motorboat, with the best of them. (Give that a google if you aren't familiar and want to see something cool.) And he holds the record for paddling around Lake Tahoe. The seventy-two-mile circumnavigation by kayak took fourteen hours and fifteen minutes. He even starred in a documentary called *The Push*, in which he faced freezing temperatures and covered the eighty miles across the South Pole on a sit-and-ski, using only his arms to propel him across the snow.

Though Grant was dealt the shittiest of hands, his way of dealing with it is to get excited about the coolest of things and then do whatever it takes to be able to do them and share them with others. For Grant especially, this takes a fair amount of foresight because, as a paraplegic, even the most elementary of everyday movements require planning and a whole lot of realistic optimism. He couldn't just go fly a plane—his feet couldn't work the pedals. He had to *invent a device* to make it even possible and then persuade the FAA to approve it for flight.

Grant doesn't simply dream about his audacious goals. He plans out each step, and when he hits obstacles, he methodically works around them. Grant's foresight and his ability to always see the positive has landed him a longstanding spot on my Dream Team.

TAKE THE 30,000-FOOT VIEW

You don't have to overcome intense physical challenges or fly jets upside down to become more resilient. Often, it's the everyday challenges that build the resilience muscle the most, because they require you to dig deep over and over again, like reps at the gym where you have to psych yourself into lifting those weights for rep number eight...nine...uh...ten! The same goes for getting your butt out of bed and off to work, summoning up a kind greeting for the office narcissist, or choosing a healthy snack when there are powdered donut holes in the break room. The more you do it, the easier it gets. If you can learn to let go of things like your frustration with traffic, bounce back from a failure at work, recover from an argument at home, or power through any number of small daily assaults, you'll strengthen your resilience even more, making it easier to turn it around again tomorrow. To start, take the 30,000-foot view, which is a way to reframe your perspective. No jet necessary. It's a three-step process:

1. **Get a new vantage point.** I often saw things differently—literally—while in the cockpit, flying 30,000 feet above earth. In an F-16, you get a 180-degree panorama from the big bubble canopy overhead that offers an unobstructed view of the sky, the clouds, and the land below, like your own personal observatory in the sky. Flipping that view upside down was even more of a perspective change.

    Up there, you can't help but feel small, and so, some of your problems begin to feel small, too. Or at least *smaller*. The 30,000-foot view makes it easier to find the silver linings in whatever trials you're experiencing.

    When the Thunderbirds were flying home on long cross-country flights, high above the ground, I could see things differently, and I'm not just talking about what was outside my jet's canopy. When the weather was good and we were in between air refuelings, I had time to evaluate where I was in life and where I wanted to go. I didn't have emails to answer or phone reception to scroll social media. It was just the here and now with nothing to do but take in the view. So, zoom out and take a new perspective. Create your own 30,000-foot view by getting away from the day-to-day. Shake up your routine. Learn something new, go for a walk, or take a hike. I've found that getting outside is one of the most effective ways to change your perspective because it has an uncanny way of making you feel small. With it, the scope of your problems feels less insurmountable. Pause from the daily distractions and take it all in. Every now and then, take it a step further and flip your view upside down, challenging your assumptions and taking on your limiting beliefs. You never know what you'll see.

2. **Acknowledge one good thing that came from the bad.** Take time to think about how you got to where you are in life and take a moment to celebrate the small wins. We're wired to seek out the bad things, so we have to work at finding the good, including the little victories. It could be the exercise program you started and then stuck with despite life's disruptions. Or it's the way you strengthened your relationship with your kids by responding instead of reacting to them. Or it's appreciating the public praise your boss gave you the other day.

    Of course, it's okay to celebrate the big wins, too. When you're taking a 30,000-foot view, it's the wins, not the losses, that help you find your silver linings.

3. **Land the plane but take something with you.** Eventually, you have to return to your everyday life. You have to land the plane. When you do, choose one thing from your 30,000-foot view to take with you. Maybe it's gratitude for a small win or a new perspective on something bad that had happened to you. It could be a realization that you want to share with your wingmen or a new understanding of an old problem that drives you to fresh solutions. Whatever it is, remember that you can always climb back up to take in the view.

THE FLIPSIDE

The three-step 30,000-foot view process provides the framework for seeking out the silver linings in everything you do. Once you know how to find it, building your resilience muscle gets easier. In fact, it's like you can't help but become resilient because your perspective has changed. It's a power reserve that you can tap into whenever you need to shift your thinking.

But there's another way to find that silver lining, only it has nothing to do with bouncing back. It's about regrets and how you avoid them and use them as fuel.

I REGRET TO INFORM YOU

When I made the decision to keep my oxygen mask on during high-altitude training, it was a choice for resilience, but it was fueled by regret—or avoiding it. Not just any regret, but a very specific type that makes up a lot of what people regret at the end of their lives: wishing they'd done something that they'd chosen not to.

You'd think that most people would regret choices and actions they've taken, but that's not the case. Of the four "core regrets" that author Daniel Pink covers in *The Power of Regret*, it's what he calls "boldness regrets" over choosing to play it safe that tend to afflict people the most.[3] In fact, in Pink's World Regret Survey, people reported regretting things they *didn't* do more than things they did by a ratio of more than three to one.[4]

Some of their regrets were not pursuing higher education, turning down opportunities to travel, and failing to connect with loved ones one last time. Not getting out of a bad marriage was also a common regret, especially among women.

This matches up with the informal findings of Bronnie Ware, a palliative care nurse and author of *The Top Five Regrets of the Dying*, which has been translated into twenty-seven languages. Among those five regrets are: *not* staying in touch with friends, *not* expressing feelings, and *not* living a life that's true to themselves. The other two are "I wish I hadn't worked so hard," and, "I wish I had let myself be happier," which are really regrets for not doing things, too, as in not making family and fun a priority and not pursuing happiness.[5]

Why does *not* doing things matter? Pink wrote, "At the heart of all boldness regrets is the thwarted possibility of growth."[6] You could have become happier, stronger, more connected, more involved, more evolved, but your choice canceled out that kind of growth, and so you stayed stuck, never knowing what was truly possible.

And why do people choose to play it safe? Fear. Fear of uncertainty. Fear of failure. Fear of rejection. It was fear, specifically fear of rejection, that initially kept me from swiping right on a guy who would turn out to be much more than a dating app match.

Shortly after arriving in Las Vegas for my Thunderbird assignment, I signed up for Bumble, the dating app where, at the time, only the woman could initiate conversation. I wrote on my profile that I was an Air Force pilot, but I didn't share that I was a fighter pilot or a Thunderbird pilot. I had matched with a guy that I thought was very attractive—so attractive that I actually thought he'd accidentally swiped right on me. The possibility of rejection and the vulnerability that came with it made me uncomfortable, and I was about to let our match time out so that it would automatically disappear from the app after twenty-four hours. I would just wait for a less intimidating guy to come along. But then I got a notification that he had used the Extend feature, paying extra to keep his profile in front of me longer.

That was just enough to bolster my confidence, to tip the scale that kept me comfortable enough to take a risk. I mustered my courage and sent him a message, and that wasn't easy for me. I can swipe right all day, but composing a clever and funny message that's alluring and interesting but not off-puttingly enthusiastic? That's a whole other thing that I imagined someone more extroverted than me could pull off.

I honestly have no idea what I said, but it must have been clever enough because he replied, and I definitely remember that message exchange:

"Hey! What do you fly in the Air Force?"

"The F-16!"

"Huh, I prefer the A-10."

Wait, what? This wasn't the reaction I was used to getting. Most people think the F-16 is pretty cool, and it usually became the main focus of first dates, for better or worse. His response caught me off guard, but it also intrigued me, and so, we went on a first date. Turns out, he had been in the Marines, and the A-10s specialize in supporting troops on the ground. During his time in Fallujah, in the most intense urban fighting the United States had seen in recent memory, the A-10 had been crucial to his unit's survival.

This guy, the one that I was too intimidated to message? Our first date, which was supposed to be just one drink for an hour, lasted five hours and included dinner. Six months after our first date, I married him. Now, John and I live together in Las Vegas with his son and a needy sheepadoodle named Scout.

When I think back on it, realistically, what was the worst-case scenario of messaging John on Bumble? I'd get ghosted? Or he'd say he's not good with technology and he'd accidentally swiped on me and then accidentally used the Extend feature and also, he's not interested? I was so afraid of rejection that I almost ended my marriage before it even began.

Regrets, I've had a few, but that's not one of them. Luckily, I used my fear of regret to power through my fear of getting rejected, and the result was an amazing partner, a new son, and a wonderful life I hadn't imagined.

Regrets can propel you to be resilient because they'll make sure you'll do just about anything to avoid them. In other words, fear of regret can motivate you to dig deep. When we do have regret, it is often "a painful experience that has some introspection," explains researcher and lecturer Marcel Zeelenberg in his TEDx talk. But, he says, "regret is there to help us make better decisions."[7] It brings us wisdom.

You might do something you regret (or regret not doing something) and it hurts. You learn from it, and next time you choose more wisely because you don't want to feel that way again. It's regret aversion, and it's when you factor in whether or not you'll regret your decision when you make it.

Sometimes it can lead you to an outstanding outcome, like it did for me on Bumble. But sometimes it can lead you toward a decision that doesn't serve you well. Take investing, for example. *Investopedia* says that regret aversion "is the tendency for people to make emotional, rather than logical decisions, in order to avoid feeling regret."[8] We drop logic in favor of emotion to avoid regret, perhaps not buying that hot stock or choosing a cheaper house you really don't like, only to regret the decision later.

Regret aversion can be good, fueling you toward growth; and it can be not so good, keeping you stuck in indecision or poor decisions, or mired in guilt and shame. Remember what I said about guilt and shame in chapter two? With guilt, we believe we did a bad thing, but with shame, we believe we *are* bad. When you've made a decision to avoid or assuage a bad feeling about your character, that usually means you're motivated by either guilt or shame. But when regret aversion is related to fear, you're essentially scared of wishing you'd done something differently.

For a long time, I regretted my behavior during my assignment in Japan. I hadn't shown up as the best version that I could have been, and I felt like I'd let down the people around me. In short, I felt guilty.

Years later when I was flying for the Thunderbirds, there was a Patch Night on base, when graduates of the USAF Weapons School, the Air Force's version of the Top Gun program, celebrate completing the program. It also draws people from squadrons all over the country and world to Nellis Air Force Base for the evening. Picture hundreds of fighter pilots celebrating and reconnecting.

That night, I ran into my commander from my final year in Japan. (Not the one who questioned whether I was a contributing member of the squadron.) It was five years after that assignment and I'd since had success as a Thunderbird, and yet, I found myself apologizing to him for my work as chief of scheduling while in Misawa. Even half a decade later, I still felt guilty about how I just went through the motions at work after my divorce. He was busy telling me how proud he was of all I'd accomplished since we worked together, but I couldn't hear it.

"I'm really sorry about how I showed up while you were the commander in Japan," I said.

"Don't worry about it. You had a lot going on." He let me off the hook. He knew I wasn't a lazy employee, and yet, I'd carried that guilt all that time. I was ashamed, and I regretted not showing up and giving it my best.

When I had left Japan for Texas, I knew I didn't want to feel that way anymore. I was capable of better work, and I made sure I rose to the occasion. I understood then that I had to feel uncomfortable at times to achieve my best because at least then I wouldn't have any regrets, even if I failed. And that was when I started to grow, both personally and professionally.

At Patch Night, my former commander gave me the grace I hadn't given myself. I'd work hard to never do that again. Regrets or not, we can all give ourselves grace when things are difficult because when we do, we open ourselves up for the possibility of growth that regrets—or avoiding them—can bring.

THE SILVER LININGS ALL AROUND YOU

The view from the canopy of my F-16 was often beautiful. Blue skies, puffy clouds, sunshine above. Or, when I flipped my jet,

THE FLIPSIDE

the landscape below: rivers, oceans, forests, deserts...I've seen it all. Sometimes we were flying so close, all I saw was another jet a few feet away. I like to think I appreciated it all, but I know that sometimes, I let the little things like the grind of the schedule, the pressure, or the time away from my family become the focus. Sometimes you have to work a little harder to find the silver lining.

If I had to sum up the "gear" you'll need to find your silver linings, it's a mix of resilience and regrets—avoiding them or learning from them. It isn't a one-and-done operation, either. It's something you keep working on over and over until it comes naturally to you most of the time.

Finding silver linings isn't a simplified process of overriding negative feelings to focus on good ones, perhaps surrounded by positivity plaques from HomeGoods. Those are fine, but seeking silver linings is more than a mentality. It's not a pep talk. It's work.

After I came so very close to yanking off my oxygen mask during high-altitude training, nearly ending my pilot career before it even started, I spent years building my resilience in tough situations. Perhaps the job required it, but my life needed it, and I became a better pilot and person for it.

When you seek your own silver linings in whatever you do, determine what will be in your way and how you will get around them. Look for things that will help you along the way. If you take a 30,000-foot view, you may see your regret steering you, your resilience fueling you, and your grace carrying you until you land. If you plan out how to reach silver linings—they're out there—the darkest times will seem brighter.

MICHELLE CURRAN

Scriptside:

- Looking only for the bad, believing there's nothing good to find.
- Keeping your world and your view of it small.
- Wallowing in shame and regret.
- Punishing yourself for being human.

Flipside:

- Looking for the silver lining every day.
- Taking the 30,000-foot view.
- Leveraging regret to avoid or learn from it.
- Giving yourself grace.

CHAPTER EIGHT
Focus on the Next Closest Alligator

I'm flying my F-16 fighter jet five hundred miles per hour over the desert in Nevada, and another jet is headed directly at me. Head-on. I've just been promoted to the Thunderbirds' Lead Solo and I'm supposed to be training the new Opposing Solo, my former position. We are performing an opposing pass, where our jets narrowly avoid each other with a space the length of a garden hose between us. We each "own" a certain side of the show line in front of where the crowd would gather below. He's supposed to fly on the outside, farther from the crowd, while I own the inside. That way, we each have an "out," a safe place to go if things don't go quite as planned.

We have spent a few flights working on these opposing passes. His tendency has been to fly wide, to pass too far from my jet. He's new. I did the same thing when I first joined the Thunderbirds. It's natural to shy away from something you're facing head-on at these speeds, and the fifty-foot spacing is measured visually, without the aid of instruments. But this time is different. This time he's flying directly at the nose of my jet, a very dangerous overcorrection.

Worse, I'm inverted. I love to fly upside down. In my role as Lead Solo, I spent more time inverted than anyone else in the formation, and it became what I was known for. But that means what was on my left is now on my right and there's nothing but brown desert floor below us, so there's nothing to visually differentiate one side from the other, his side from mine. The F-16 is a small aircraft; pointed straight at you, it appears even smaller. So, as we get closer with a combined one thousand miles per hour of closure, I wouldn't be able to tell that he hasn't set the appropriate offset between us until seconds before we would collide. I'm disoriented. And he's still flying right at me.

I have a split second to make a decision. In a situation like this, I should move to my side. That's what we're trained to do, and that's what I'm training him to do. But I can't be sure which side is my side and it would be disastrous to choose the wrong one. This is all happening so quickly that once I recognize he's too close, I don't even have time to open my mouth and key the mic to say "Five is aborting" about me, or "Six, abort" at him. There's nothing else to consider, no alternative plan to attempt. I have to focus on a single thing: trusting that his own self-preservation instincts will kick in and he will fly around me. So I override the urge to make some sort of correction and instead, I hold steady, at five hundred miles per hour, a few hundred feet off the ground, upside down, and I wait. And by wait, I mean mere seconds pass from the moment I realize that he's too close to possible impact. But it feels much longer. Time slows down and I hold my breath.

At the last second, he rolls up his wing and passes over my jet, narrowly avoiding a catastrophic collision. I'm inverted so I can't see him at this crucial moment. All I can see is that his jet starts to roll as he disappears below... or above?... the belly of my aircraft.

THE FLIPSIDE

Later, when we watch the cockpit tapes, I can see just how close his jet was as it passed my front windscreen, and it's closer than anything I'd seen before or since.

At that speed, the fifty feet of spacing we usually targeted for these passes always felt pretty darn close, allowing us to see details on the other aircraft as it passed. But what stands out for me this time is just how big his jet seems. We avoid disaster by mere feet, possibly inches. I'll never know for sure, but I'll always remember this feeling.

In the cockpit after the near miss, there is a long silence. I'm taking deep breaths, recovering from the shock of a dangerously close call. I'm angry. I'm scared. After missing two radio calls, I finally call for us to turn our smoke off and clear the show line. Once my jet is no longer upside down or near the ground, I notice the physical feeling of an adrenaline spike, increased heart rate, dry mouth, shaky hands. I consider if I should call out the mistake on the radio right then and there or if we should just knock it off and go home. I know my feelings of anger and fear don't serve me well in the cockpit, and so I regroup. We don't have any more head-on passes planned for this practice, and the discussion of what just happened can wait until we are on the ground at zero knots and one G. I make the call to set up for our next maneuver.

In the cockpit and in life, how we choose to respond to difficult situations determines our success or failure. But sometimes, it takes a near miss to remind us of our ownership of our own responses. In this case, I was forced to focus on what absolutely needed to be handled right away because I didn't have the luxury of time where I could plot, plan, or ruminate, and ultimately get stuck in the weeds. But it took training and habit building to get to the place where my instincts automatically kicked in, leading me to prioritize correctly. My trainee hadn't yet reached that level of expertise with this type of flying, so he was focused on closing the gap between us to achieve a "good hit," when the jets appear to pass through one another, eliciting a gasp from the crowd, making an execution error that's not uncommon when you're new and task-saturated. People in new and unfamiliar situations can make mistakes, and this one was extremely dangerous.

Even as a seasoned Thunderbird, I fell back on one of the earliest but most important pieces of advice I ever received in my flying career: "Focus on the next closest alligator to the boat." It means that tackling big goals or learning difficult skills will feel overwhelming, if not terrifying, but if you focus on the most crucial step you can take at this moment, you can save your ass.

I learned the phrase on day one of pilot training. It was a formal release day, which means you had to be at the office or at the squadron in the flight room for twelve hours whether you were flying or not. Even if you finished all your work, you couldn't leave until you were released because leadership wanted the team to be studying and bonding. They even ordered us to wear our blues uniform instead of the comfy green pajama-like flight suits we fly in. These uniforms required ironing, and some people even wore garters to clip the bottoms of their shirts to a band around their feet, which felt like walking around with resistance bands running down your legs all day long, but at least your shirt looked sharp.

Our flight commander did the intro, and it sounded like a scene from any Hollywood military movie: "Good morning, ladies and gentlemen and welcome to Undergraduate Pilot Training. You are the best of the best, blah, blah blah..." Then they warned us how hard the year was going to be, and how not all of us would get the aircraft we'd come there to fly. Some would wash out, meaning we'd quit or we'd be sent home.

It was super stressful because for many of us, becoming a pilot was a major dream, one that some of us had had since childhood. It was all we wanted to be and do. Our flight commander knew this, so he was preparing us for the overwhelm we would invariably feel and warning us not to let it snowball out of control.

"If you focus on the entire year, you'll get discouraged and you're going to want to quit," he told us, adding a warning about what would happen if we failed a flight and they noted it, in glaring red marker, in our grade books. You didn't want to "bust a ride" or "hook a flight," the slang for failing, because it meant that you hadn't met the standards, which varied depending on the training phase, for the task at hand. Early in the program, the standards we missed could be as basic as not taxiing on the centerline or landing too far down the runway. Later, they'd include tasks like flying formation, aerobatics, and instrument approaches. A hooked flight meant you'd have to take that ride again and demonstrate proficiency. If you failed repeatedly, you would quickly find yourself headed down the path of a series of progress rides, flights where you were being evaluated to determine if you should be allowed to stay in the program. "If you bust a ride, it may feel like it's all over, but it's common to feel this way. So focus on the next closest alligator to the boat."

In other words, take on the most immediate threat—the next flight—and worry about the rest later. It's essentially a reminder about the difference between pressure and stress. To me, pressure is usually positive, while stress often is not. Pressure can push you to perform better through urgency that motivates by increasing excitement. It's related to a short-term situation that requires immediate attention. Stress, on the other hand, is typically related to a longer-term situation, and it can trigger negative emotions, such as fear or anxiety.

The more advanced your skill level, the more likely you'll feel pressure instead of stress. By the time I was looking at an F-16's nose pointed right at mine, I'd been in the Thunderbirds for more than a year, so my skill level was greater than the other pilot's, who, though a seasoned fighter pilot, was new to the unique skills required of our squadron. But I'd only recently become an instructor pilot for the Thunderbirds, so monitoring another pilot in this environment was new to me. That part was stressful. It would have been easy to succumb to the overwhelm of all I had ahead of me—how much I still had to teach the new solo pilot and finding my own footing as an instructor. But the next closest alligator was finishing the flight safely and getting the most out of the training time. The important thing in these situations is to avoid flooding your own system with sensory overload.

GETTING INTO FLOW WITH THE YERKES-DODSON LAW

Medical News Today explains that sensory overload is when the brain receives too much information at once, overstimulating one or more of the body's five senses, leading to anxiety, fear, agitation, overwhelm, irritability, a loss of focus, and stress, among others.[1] It can feel worse when you're tired or hungry, which explains the special overwhelm that is being "hangry."

But what if there was a way to maximize your performance even in times of stress? One way is to apply the Yerkes-Dodson law, a psychological principle that can help you identify the optimal level of "arousal," which in this case means the state of being alert, awake, and attentive, to achieve peak performance.[2] Created by two researchers in the early twentieth century, the law states that when arousal is very low or very high—boredom or anxiety—performance tends to suffer. But there's a sweet spot in between to aim for.

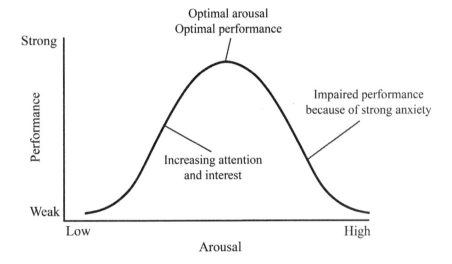

On the way up the U-shaped curve, your attention and interest are increasing, and you're beginning to feel fired up and excited, like right before a tennis match or when the roller coaster is click-clacking up the track. On the other side of the upside-down U, your performance becomes impaired because of your anxiety. So impaired that you're unable to process things like hearing simple instructions.

That was exactly what happened to me during an air-to-air large force exercise, a sort of war game for fifty-plus aircraft of all different types that's among the culminating events of the year-long F-16 basic training course. In my first-ever run, my helmet was on fire. Not literally, of course. It's what we call that feeling every fighter pilot can relate to: It's as though you're moving and thinking in slow motion, and always a little too late, like your brain is working so hard you wouldn't be surprised if smoke was coming out of your ears. In the cockpit, that can rapidly narrow your situational awareness to the point that it becomes dangerous.

A large force exercise is a lot like playing defense in a high-stakes soccer game, in three dimensions, at four hundred miles per hour and thousands of feet in the air. I was on the "good guys" side, and our job was to protect the base, "our goal." We did this by "shooting" down the "bad guys" aircraft before they could get close enough to release their simulated bombs and strike our base. I understood my role, but my brain was having a hard time keeping up with the combination of complex tasks that, until then, I'd only read about and performed in a simulator. At any given moment, I had to monitor my own airspeed and altitude, keep track of my flight lead, listen and respond to radio calls, complete the pre–weapons release checklist, be aware of my range from the target, and more. To me, F-16 training felt like drinking from a fire hose—so much material to read and so many skills to master all at once.

In that moment, I couldn't process the radio calls as I was simultaneously flying my F-16 and monitoring my radar to find the bad guys. I was supposed to "shoot" a missile at them soon after detecting them with my radar. Not long after shooting the missile, I would turn and leave to avoid being shot by the enemy missiles that were surely headed back our way, but only after I heard a particular radio call. But I didn't hear it, so I didn't leave when I was supposed to.

At the debrief later, my instructor asked me why I didn't stay on the timeline to turn around. In my mind, the radio call was either too quiet or it didn't come at all. But we listened to my tapes and there it was, clear as day: the radio call I didn't hear because I was too task-saturated to take it in.

My brain was like an overheating engine, a classic case of sensory overload, which meant my system stopped processing some visual and audible cues. It's like when you're driving and you can't find where you're supposed to be going, so you turn down the radio because it's hard to focus when you're distracted by Lizzo telling you that she's feeling good as hell.

I didn't yet have the experience to make decisions using the big picture without hearing or seeing very specific things to prompt me. I would have to learn to compartmentalize just enough to get the job done, but that would take more training and learning to trust my own abilities when the pressure was on. I'd have to learn to operate under high stress and eventually use it to my advantage. Until then, I'd continue to experience helmet fires.

On the Yerkes-Dodson curve, performance reaches its optimal level somewhere in the middle. If you stay on the left side of the curve, low stress can bring on lack of motivation and complacency. The high arousal on the other side of the curve is where you can end up in fight, flight, or freeze, and my response in that training was essentially to freeze.

Psychology Today defines the three states of the Yerkes-Dodson law curve as disengagement, frazzle, and flow.[3] In disengagement, you're uninspired and disinterested. In frazzle, you're too overwhelmed to get much done. But in flow, you're able to "channel positive emotions to an energized pursuit of the task at hand." Had I been in flow during the air-to-air large force exercise, I would have been able to hear that radio call and complete my mission.

But how can you stay alert without getting overwhelmed? By

focusing on the next closest alligator to the boat. This simple suggestion became a mantra that allowed me to excel in the program. While there are a lot of alligators that fighter pilots have to focus on in training and beyond, I've since learned to apply this mantra to other parts of my life. Whatever your big audacious goal is, break it down into smaller steps and your system won't get flooded in times of stress.

For instance, if you've ever run a marathon, you understand that you don't just go out and run 26.2 miles to start. You break down that big goal into little pieces: Start by running a few miles, maybe racing in a 5K and then perhaps a 10K or a half-marathon. Even when you're in the marathon on race day, you're not thinking about mile twenty-six when you're two miles in. You're thinking about the next closest alligator—the next goal to achieve, such as passing that runner who's just ahead of you, getting to the next aid station, or hitting five miles and then ten. That doesn't mean you lose track of your why or your goal. You can still be motivated by a big goal while focusing on the next alligator.

THE SELF-EFFICACY SOLUTION

When it comes to getting into flow and focusing on the task at hand, self-efficacy is important. It's your level of confidence in your ability to achieve your goals. That may sound like self-confidence, and they are related. But self-efficacy is about the *likelihood* of accomplishing tasks based on four key components, as introduced by Stanford psychology professor Albert Bandura and studied for decades since.[4]

THE FLIPSIDE

1. **Vicarious experiences:** Observing when role models successfully complete a task can influence your belief in your own mastery. For instance, when I saw others in my squadron successfully complete a difficult upgrade program to become an instructor, I felt like I might be able to do it, too.

2. **Emotional arousal:** Emotions and bodily sensations affect your belief in your own abilities. If you feel excited to try something new, you are more likely to accomplish it. Reframing anxiety into excitement is a useful tool to generate this feedback, as we discussed with Dream Teams.

3. **Performance accomplishment:** Prior success and personal experience can affect how competent you feel at similar tasks. The second time I ever flipped my F-16 over to fly upside down in Thunderbird training, I felt more self-efficacy because I'd done it successfully before. Whenever you have a win, celebrate it, and set yourself up to have a whole slew of positive performance outcomes to pull from.

4. **Verbal persuasion:** Encouragement can affect your ability to perform. But verbal persuasion doesn't have to come just from other people. It also comes from your own inner critic. Leaning on your wingmen and gaining perspective on your inner critic are critical to craft a positive verbal persuasion.

Self-efficacy can also help us face our fears. Take snakes, for example. In 1982, Bandura set out to test the effects of self-efficacy on overcoming phobias—in this case, a fear of snakes, otherwise known as ophidiophobia. In his study, one group of people with ophidiophobia would interact with snakes, holding them or letting them crawl on them (who signs up for these studies?), while the other group observed. Then Bandura assessed each group's fear of snakes after the experiment and found that the snake-interacting group demonstrated higher self-efficacy and a lesser likelihood of future avoidance than the group that had merely watched.[5] By exposing subjects to their fears, it was shown to help people to better cope with them.

Bandura's study demonstrated that one path to achieving self-efficacy is enactive self-mastery, which is essentially learning by doing. Bandura says self-mastery can "provide the most authentic evidence of whether one can muster whatever it takes to succeed."[6] But winning isn't enough. He says that we have to recognize what it took to win.

For example, think about what it takes to learn to drive a car. When you started driving, you might have known a little bit about how to steer and brake by watching your parents, older siblings, or friends drive, but until you got behind the wheel, you couldn't achieve self-mastery. After you drove a few times without crashing the car, though, you began to master the task, thereby developing self-efficacy. You proved to yourself you could do it.

There's even collective efficacy, which is a group's shared belief in its joint capabilities, and that is surely very important in any squadron, but especially in one that operates with the kind of precision required in the Thunderbirds. But for any human, how we think, feel, and act affects our confidence in our ability to perform as a unit.

In the military, this is called "cohesion," and it's defined as "the 'cement' binding together group members and maintaining the relationships to one another."[7] You can't be combat-ready if you're not a cohesive group with collective efficacy. In any squadron, there's a correlation between cohesiveness and morale, and I would venture to say it's the same thing with individuals. If you're feeling "good as hell," you feel more pulled together and, perhaps, you're more likely to feel self-efficacy. You're more likely to believe in yourself.

It's fair to say that the pilot involved in our near miss proved our cohesion that day, avoiding a disaster and falling back on the training he was receiving, the trust between us, and the belief that we were both doing our best to work toward a common mission. This allowed us to continue to train and grow together for the months to come.

Throughout my career, I kept challenging myself, moving from ROTC to active duty, from pilot training to the F-16, from a pilot to an instructor pilot, from combat missions to flying upside down, mere feet from another jet. Each time, I had to rebuild my self-efficacy, first by learning and then by doing. Lucky for me, there were no snakes involved.

If you're the person who's less likely to raise your hand to volunteer for new things, your reluctance can affect your self-efficacy, which can affect your self-confidence, ultimately influencing whether or not you take on a growth mindset. (See chapter five.) It all comes down to believing in yourself. The more I proved myself in the cockpit, the more self-confident I became and the more self-efficacy I built.

DON'T LET DECISION FATIGUE DRAIN YOU

All the training that Air Force pilots go through is set up to break down difficult tasks into small steps so that we can learn self-mastery one alligator at a time. I might have missed an important verbal cue in the air-to-air large force exercise, but eventually, I would become good at flying the F-16. It's just that there was a lot more yet to learn and plenty of training ahead of me.

With all those tasks and lessons, I'd face my share of decision fatigue, which is when the quality of your decision-making declines after making lots of choices in a row. During a single flight over just a few minutes, you'd need to decide to adjust your throttle and your stick inputs, determine when to speak up on the radio, and decide what to tackle first, all while completing a laundry list of other tasks. By the time you made a single radio call, you'd have to once again assess your flight parameters and adjust the stick and throttle. Done with that? Well, now you're overdue for a checklist to be completed before employing ordinance. There were always several tasks that felt both urgent and important, and it was up to each pilot to decide how to manage them.

Decision fatigue can tire you out and leave you with brain fog. Psychiatrist Lisa MacLean, MD, told the American Medical Association that we make, on average, some thirty-five thousand decisions in a day, and that can deplete us. "The more choices you have to make, the more it can wear on your brain, and it may cause your brain to look for shortcuts," she explained, adding that there are four main symptoms of decision fatigue:

1. procrastination
2. impulsivity
3. avoidance
4. indecision[8]

THE FLIPSIDE

The pandemic exacerbated decision fatigue for a lot of us because it added many more, often complex, decisions to our plates. In fact, a 2021 survey by the American Psychological Association[9] found that about a third of adults in the United States struggled to make even basic decisions about simple things like what to wear. My peers, the Millennials who were born between 1980 and 1995, had it hard, with nearly half reporting difficulty with decision fatigue. Parents with kids under age four had it hardest of all, with 54 percent reporting stress from daily decisions.[10]

This is where establishing habits can save the day—and your energy—and it's the reason why military training is so regimented. Part of it is to establish discipline, but it's also because good habits can reduce decision fatigue. If you know you're supposed to wake up every morning at oh six hundred and immediately make your bed—there's no decision to make. In fact, retired U.S. Navy admiral and former Navy SEAL William H. McRaven wrote a book called, *Make Your Bed: Little Things That Can Change Your Life... and Maybe the World*. He considers making your bed each morning an accomplishment of the first task of the day. Even better, you don't have to decide whether or when to do it if it's a morning ritual, a habit.

One of my friends, Shayla, uses her to-do list to form habits and reduce decision fatigue. A very successful mortgage broker in Reno, she worked with her team to close over $2 billion in business. To achieve that kind of success, Shayla makes a to-do list that helps her feel like she's making progress.

Instead of twenty things to cross off, her list has six items every day, which she prioritizes. If she only gets to four of the six, she moves two to the next day. Anything that's carried over for three days gets scrutinized. She asks herself, "Do I really need to do this? Is it that important?"

Then she either delegates it to someone else or enlists help to accomplish it. Sometimes, she takes it off the list altogether. That way, she doesn't go on autopilot, rolling over a to-do item over and over again, while still giving herself the grace to not complete her tasks if needed. It's good habit making because it's realistic and yet effective. It's also a great example of a contract with herself that balances discipline and grace.

When you establish habits, you can save your mental energy for bigger decisions—and more alligators. If I decide on Sunday that I'm going to the gym on Monday, Wednesday, and Friday that week, then I don't have to expend the mental energy to figure out when I'll go or to motivate myself to exercise. You can even wear the same thing every day, like Steve Jobs did, or eat the same thing for breakfast each morning. You can establish a bedtime for yourself. (Hey, it works for the kids!) And you can add habits that make your life better or your body healthier, like drinking a certain amount of water each day or always taking the stairs instead of the elevator at work.

Good habits prevent you from having to use willpower, which James Clear, the *New York Times* bestselling author of *Atomic Habits*, says is like a muscle. It can get fatigued by overuse.

"When your willpower is fading and your brain is tired of making decisions, it's easier just to say no," he wrote.[11] Establishing habits so that willpower doesn't have to do the work is one way to reduce decision fatigue. My friend's to-do list doesn't require willpower to complete because she relies on other parameters to complete it, including simplifying choices, asking for help, and taking breaks.

THE FLIPSIDE

Your habits reduce decision fatigue, which in turn helps you accomplish your goals. It's one way to focus on the next closest alligator. Another way is to reframe how you're thinking of them.

COGNITIVE REAPPRAISAL

Reframing starts with cognitive reappraisal, a reinterpretation of your situation. It's got a touch of seeking silver linings and a dab of overcoming your inner critic to it, and employing emotional arousal to your benefit, but it's even more involved than that. In cognitive reappraisal, you're actively reframing a negative situation as a positive one.

Simon Sinek did it when he told himself he was excited, not nervous, before delivering an important speech, but surprisingly, it's not a common practice. The American Psychological Association[12] reported that people don't do it that often because it's difficult work. Instead, people tend to fall back on different methods for emotional regulation in light of bad news via distraction, such as online shopping, doom-scrolling, or lying around watching the big game, or through suppression, as in a fake smile or a stiff upper lip. But those aren't forms of cognitive reappraisal. They're Band-Aids.

True cognitive reappraisal, which is often used in cognitive behavioral therapy, is a way to rethink your perception of a stressor or situation. For example, instead of thinking that the train delays you're experiencing are a stressful inconvenience, consider it a chance to go grab that latte you've been craving or to listen to a podcast and people-watch. I mean, you can't make your train arrive any faster, so why get all worked up over it? It might even keep you from sliding down the Yerkes-Dodson curve into anxiety.

You can use cognitive reappraisal for major life events, but that takes more finesse and fortitude, and perhaps some therapy. One of the ways to begin to incorporate cognitive reappraisal is to watch for "thinking traps." This can include:

- **All-or-nothing thinking:** When something is either good or bad or a success or a failure, and there are no in-betweens. ("I'm a failure!" instead of, "I haven't figured it out yet.") Many of the most important alligators live in the gray area.
- **Catastrophizing:** Where you make assumptions about how the situation will play out and it's always bad. Examples are: "I'm going to get fired," "My spouse will divorce me," or "I'll die penniless and alone."
- **Should-ing:** When you let the way you think how the world ought to be or you ought to be skew your perception of reality. Bottom line: Don't "should" all over yourself.[13]

Cognitive reappraisal can help you reframe the negative and enhance the positive to improve your well-being. It's a matter of editing your negative self-talk, watching for thinking traps, and looking at things in a more positive light.

I've introduced a lot of concepts for a chapter about keeping yourself from getting overwhelmed, which I hope hasn't been...well, overwhelming! But when put together they provide a framework for focusing on the next closest alligator to the boat without letting the stress of the situation distract you.

First, you take on the most immediate threat—the pressure—and leave the rest (the stress) for later. Using the Yerkes-Dodson law can help you do this, allowing you to find the sweet spot between boredom and anxiety where you can achieve optimal performance.

THE FLIPSIDE

Offload tasks when you find yourself on the backside of the curve. Then, use the elements of self-mastery to reach self-efficacy. In other words, learn by doing so that you show yourself that yes you can. (Snakes optional.) Avoid decision fatigue by breaking down big tasks into smaller ones—you know, that alligator closest to you. Finally, reinterpret your situation through cognitive reappraisal. It's a way to give yourself the grace, time, and skills you need to succeed and to make the immediate pressure feel less stressful. If you focus on the next closest alligator to the boat instead of allowing yourself to be overwhelmed, you can stay calm and cool no matter what's headed right at you.

Scriptside:

- Trying to take it all on at once.
- Falling victim to boredom or stress.
- Failing to build self-confidence through mastery.
- Letting bad habits overwhelm your decision-making.
- Thinking like Eeyore: It's always a bad day.

Flipside:

- Focusing on the immediate problem to solve.
- Finding flow on the Yerkes-Dodson law curve.
- Facing your fears by building self-efficacy.
- Reducing decision fatigue by forming good habits.
- Using cognitive reappraisal to rethink the negative.

CHAPTER NINE
You're Better Lucky Than Good

My wingman and I are watching our target, a known path for terrorists in Afghanistan. We make up a two-aircraft formation tasked with completing this mission on a clear July night just a couple of weeks into my first deployment. We fly just a thousand vertical feet apart in a wheel formation, in the pitch-black of night. As if being in combat wasn't making me nervous enough, such a close flight path makes the threat of one jet hitting the other a real possibility.

We've dropped our laser-guided bombs and they're on their way to the target. I can tell, because a symbol is flashing on the small screen that displays my targeting pod, which is an infrared laser system that guides the bomb. I should be concentrating on the outcome—taking out the Taliban's ground route—but instead, I have this overwhelming feeling that I need to confirm where my flight lead is to ensure that we aren't going to hit one another.

I glance at my altitude and then quickly check another screen that tells me he's a safe distance away—all in a split second. But when I return my attention to the small screen that displays my targeting pod for monitoring the bomb, the symbol that shows my laser is firing has stopped. The bomb is no longer being guided, and it hasn't yet hit the ground. Oh no. My stomach drops.

My finger's still pulling the trigger that fires the laser, but I must have eased off on the pressure just enough for it to stop. By this point, I haven't dropped many live laser-guided bombs, which are precise when you use them as directed, but keep going on whatever trajectory they were on when they lose their laser guidance. As soon as I saw the laser had stopped firing, I immediately squeezed the trigger harder and the flashing resumed, but now, I don't know for sure where the bomb will land.

I'm looking through my targeting pod and I'm watching the countdown to impact. As the timer hits zero, I don't see the bomb detonate in the narrow field of view of the pod, and a terrible, sinking feeling comes over me. I quickly zoom out: It landed in a nearby field. It could have had devastating results, but in this case, the bomb landed in an uninhabited area. Yet I feel the weight of how serious the outcome of this mistake could have been.

My flight lead's bomb was able to take out the Taliban's path and complete our mission, but I feel terrible. After we land back on base an hour later, the director of operations, a seasoned pilot and weapons school instructor who's our squadron's second-in-command, meets us in the debrief room. We watch the tapes together, and he points to when the laser stopped firing.

"What were you doing at that point?" he asks me.

"I misprioritized." I explain that I'd chosen to "deconflict"—check my location in relation to the other jet—rather than keeping my eye on the laser. I was dropping a live weapon in a deployed environment, and yet I felt like something else was more important.

He talks about how the attack was set up and suggests other ways we could have done it so there would be zero concern about deconfliction, and by the time we're done, I know how to avoid it in the future. He could have easily shamed me, leaving me to walk away defeated, questioning my abilities as a fighter pilot. Instead, he reminds me of the role luck played that night, how I didn't hurt anyone or change the outcome of the mission. In the end, it's a best-case scenario of what could have been a very grave situation.

"You'll fly again tomorrow." And then, he never brings it up again. No one does.

It's a classic example of the fighter pilot expression "Better lucky than good." We meant it ironically because of course it's better not to rely on luck in the cockpit. In our debriefs, the vast majority of the time, luck was not among the root causes of the outcome. Usually it was an identifiable issue, such as a lack of information, a wrong decision, or low experience level. In my case, it was a misprioritization exacerbated by the fact that I had a lack of proficiency with this weapon in the stressful environment that is a combat setting.

I did a dumb thing. But after that night, I never did it again. And I learned that we make our own luck.

THE CASE FOR SMART LUCK

There were plenty of times in my career as a fighter pilot where dumb luck saved the day. Just like that night over the target in

THE FLIPSIDE

Afghanistan, random chance worked in my favor. But what if you could create the circumstances that help good luck find you in the first place? What if there's "smart luck"?

A Wharton School professor thinks there is. Paul J. H. Schoemaker, PhD, published a study in 2021 with "strategies to get Lady Fortune on your side" through various approaches designed to improve your odds of a favorable outcome.[1] Though his research focuses on business, the overlap where luck and preparation meet can apply broadly to all our lives. In other words, your meticulous planning can draw good luck your way. Some of his strategies for creating smart luck are:

Position Yourself Better: Be where lucky breaks seem to happen. This is why aspiring actors move to Los Angeles and New York, and fledging country songwriters and musicians move to Nashville. They're more likely to catch their big breaks where the decision-makers in their businesses are located. These days, you can get discovered on TikTok or YouTube, like how I unexpectedly got my first speaking gig thanks to posting about the Thunderbirds on Instagram, but if you don't have content that you're constantly feeding online, you won't be positioned for smart luck to find you.

Improve Peripheral Vision: Focus, but don't be so single-minded that you miss other opportunities when they come along. When you get stuck on the idea that your goal or dream can happen just one way, you might miss opportunities that had never crossed your mind. As Schoemaker wrote, "Don't be the person who looks for their lost keys only near the lamp post, where the light is best; search a wider field, explore the shadows, and hone your peripheral vision."

In other words, don't get stuck in the drool bucket. That's what we called the cockpit displays, screens, and instruments that can become the sole focus of new pilots during stressful moments. Imagine you're using your radar to lock up a bad guy who's tens of miles away. You can't see him with your eyeballs just yet, so you're using your radar to put your aircraft in the perfect position to intercept. But there comes a point when it's easier to just look up, because on a blue sky day, you can easily see the other aircraft while it's still miles away. Instead, your eyes are locked onto a screen until your instructor says, "Bro, look outside."

Adopt a Portfolio View: Don't think about risks as isolated ventures because, over time, one big win can pay off many small risks. You likely already know about spreading your investment risk with a diversified portfolio of stocks, bonds, and real estate, and that's useful advice. But this is really about statistics and at bats more than anything else. It's when you refrain from putting all your eggs in one basket and also take more shots on goal, to mix metaphors.

Every choice I made that put me in a cockpit, from pilot training through the Thunderbirds, was a risk. Without those choices, I wouldn't have had my Air Force career, and I wouldn't be delivering keynote speeches or writing this book. Taking a portfolio view can bring the payoff when luck comes knocking on your door.

That should include a portfolio view of your friends and colleagues. Put yourself in the company of other people who are also willing to operate with courage and zest for life, leading to what I call pinch-me moments, risky but ultimately fulfilling adventures. For me, traveling to Argentina and paragliding over the foothills of the Andes with a fellow former Thunderbird, climbing a cliff face for more than one thousand feet on the towering red rocks outside of Las Vegas with a Navy SEAL, and wake surfing on the aqua blue waters of Lake Tahoe with my buddy Grant are all examples of pinch-me moments. You can also have these moments with business partners, classmates, your fellow Little League parents, the Polar Plunge crew—anyone who shares your quest for a goal.

Take Calculated Risks: From the outside, it appears that I have done many high-risk things that have paid off. But when it comes to taking calculated risks, I learned quickly that you have to be highly aware of what you're willing to lose and then choose accordingly. Now that I'm out of the cockpit, I tend to assess risks as "Well, what's the worst thing that could happen?" Usually, the worst-case scenario is unlikely and it's usually just a fleeting moment, a hiccup I can learn from.

When you take calculated risks, you're making informed decisions while minimizing the potential downsides. General George Patton said, "Take calculated risks. That is quite different from being rash." He ought to know. In 1944, he led Operation Fortitude, a military maneuver designed to deceive the Germans using decoys, fake radio signals, phantom field armies, and counterintelligence agents. As a result, the Germans thought the Allies would invade Pas-de-Calais in France when really they were planning the invasion at Normandy. D-Day.

It was so successful, Hitler thought that the Normandy invasion was the deception and held back his Panzer units. By the end of the day, some 155,000 Allied troops had come ashore, taking control of eighty square miles of the French coast. Less than a year later, the Germans surrendered and the war ended.

Your calculated risks don't have to be quite as complex as Patton's, of course. But you'll need to take a few steps to follow his lead. First, get out of your comfort zone or you'll stay right where you are. Embrace uncertainty but hedge your bets. For example, if you want to change your career (I have!), it's going to feel weird at first. But if you learn all you can about that career, interviewing people in the field, taking any necessary courses and/or tests, reading books, and even imagining yourself in that role, taking a calculated risk will feel less scary. Finally, be creative and open to new experiences. The road to your goals might take a detour or two. Follow them. You never know who you'll meet or what you'll learn along the way, and the positive experiences will be worth it. *Harvard Business Review* examined the "five dimensions of curiosity," as a model for generating positive experiences.[2] The dimensions are:

- **Deprivation sensitivity:** Finding relief by filling in a gap in knowledge. You google things like, "What year did Pink release 'Get the Party Started'?" or "Why do trains sound horns in the same spots?" just because you want to know. (FYI: 2001, and trains are required to sound their horns at railroad crossings.)
- **Joyous exploration:** Seeking wonder in the world. You're the one watching videos of the northern lights or a live feed of a nest of baby eagles.
- **Social curiosity:** Interest in what others think and do. You ask people about their opinions and where they got that cool coat.
- **Stress tolerance:** Exploring new things even if it's uncomfortable. You're the one signing up for whitewater paddle boarding for the first time, and you keep getting back up on the board no matter how many times you fall off of it.

- **Thrill seeking:** Risk-taking for the sake of excitement. Jumping out of a perfectly good airplane, which some call skydiving, is counterintuitive to most pilots, but I've done it and it was quite exciting.[3]

LUCK MATTERS

Is it really better to be lucky than good? Naturally, there's *privilege*, where your upbringing or circumstances provide rights or immunities you might not otherwise have had. Luck, on the other hand, is usually thought of as random, a sprinkling of fairy dust at the perfect time. But is it? You can imagine how frustrating it is for me when people tell me, "You're a lucky girl!" First of all, do they say, "You're a lucky boy" to my fellow fighter pilots and Thunderbirds? I doubt it.

Let's look at that luck they're attributing to me. I'm lucky I didn't come of age before women were finally given permission to fly fighter jets in the military in 1993. Yet even now, three decades later, if you walked into a room full of Air Force fighter pilots, just five out of one hundred would be women. Thirty years for 5 percent.

Our fighter jets were designed to meet the specifications of the average male pilot, so I guess I'm lucky I'm not five foot two.[4] And our flight suits, as I've said, weren't designed for our bodies, though there is technology being improved that will allow female fighter pilots to pee into a cup sewn right in and attached to a urine removal system.[5] Lucky them!

If preparation, opportunity, persistence, and believing in yourself make up luck, then, sure, I've been lucky. There's more to "right place, right time" than just showing up. Some people may be lucky enough to be discovered by a talent agent while buying tangerines at the grocery store, but that's not what you should count on. Put in the work, be persistent, and be prepared, and luck will find you.

One way I get ready for luck to find me is to normalize fear and to champion courage. Doing new things is scary, so when fear shows up, say hello and welcome it. Go ahead and acknowledge the worst-case scenario. Most times, the reward outweighs the risk. If that's so for your situation, then summon the courage to move forward by using regret as a motivator, asking yourself, "How will I feel six months from now if I don't do this?" Often, I realize I would regret not doing the new thing, and this becomes enough to motivate me to give something a try that I might not have if I'd let my inner critic have at it.

But why isn't this always enough? Why can't you just get by on hard work and talent alone? One of my favorite authors, Ryan Holiday, studies the Stoic philosophy, which asserts that we can control only our own thoughts and actions. He reminds us that the world can be a random place, explaining that we can do hard work, putting forth our best effort every day, but it's not up to us whether that work will be rewarded.

"Do the work. Be happy with that. Everything else is irrelevant," he wrote.[6] That's still preparation meeting opportunity. And doing the work has value in itself. I mean, why not be ready just in case luck shines on you? But if you're aware that hard work and talent don't always bring the results you hope for, you'll suffer less disappointment.

In the Air Force, we prepare through "chair flying." It's a way

THE FLIPSIDE

to get your brain up to speed when you're new in the cockpit or out of practice for something you need to do in the air. Chair flying is just what it sounds like: We can't always get time in the simulator or flight time, so we pretend to fly from a chair.

When you are a beginner, you start with a poster of the cockpit hung up in a room. I'd take a seat in the chair and position my body just like I would in the jet. I'd say my actions out loud, announce radio calls, and ground taxi instructions, practicing for my time in the aircraft. Sometimes, I'd close my eyes and visualize the flight, and sometimes, I'd reach out to "flip" a switch on the poster just like I was in the cockpit.

I'd even pair up with another pilot so we could give each other steps and test each other on what we'd do in various situations. That way, when we got in front of our class and under evaluation by our instructors, it was easier to apply what we'd practiced to stressful situations.

It may seem silly to picture grown pilots playing pretend in chairs, but there's science behind it. When you practice, you build up layers of "insulation" around the nerve fibers in the brain and increase how fast your brain can recall, command, and respond.[7] Practice may not make perfect, but it can help improve your performance, and that was our goal.

Chair flying helped us improve our skills, and by getting better at what we did, we made it easier for luck to find us. Two Italian physicists set out to quantify the role of luck and talent in successful careers, believing that Western cultures tend to underestimate the importance of external forces on success.

What they found was that talent wasn't irrelevant, but "in general, those with greater talent had a higher probability of increasing their success by exploiting the possibilities offered by luck."[8] Talented people make luck work for them.

CREATING SERENDIPITY

You can make luck work for you using some of the things we've already talked about in this book: promoting a growth mindset, challenging your inner critic, allowing yourself the grace to be a beginner, responding instead of reacting. But one of the most useful ways to exploit luck is through serendipity, the ability to find value in the unexpected and use it to create positive outcomes. As Pulitzer Prize–winning playwright Katori Hall said, "Serendipity always rewards the prepared."

Serendipity is often considered chance or luck, and it's a fairly modern idea. In fact, if you google "serendipity" and scroll past the ice cream shops and a moderately ranked 2001 movie starring John Cusack, you'll find that the usage of the word didn't really start to take off until the late twentieth century,[9] and frequently in stories about the accidental discovery of antihypertensive drug Viagra's, um, side effects.

Mark de Rond, PhD, a professor at Cambridge Judge Business School, wrote in "The Structure of Serendipity" that breakthrough innovations are "made by individuals able to see bridges where others saw holes."[10] Serendipity is what we make out of seemingly random occurrences. It's an ability, not an event.

Dr. Christian Busch spent a decade studying how unexpected moments can work in our favor after he had a near-fatal car accident that left him questioning the meaning of life. The author of

THE FLIPSIDE

The Serendipity Mindset, he said it's not enough to have an unexpected event happen, but we have to have the tenacity to do something about it. He warned that we can miss the serendipity if we don't see it or if we don't connect it to a positive outcome.[11]

But how do you connect the dots? Busch said you can cultivate serendipity in several ways:

1. **Ask open-ended questions.** It gives people leeway to bring up something you may not have considered. Sometimes, someone in a different field or with a different perspective surprises you with an idea you couldn't have thought of on your own.

2. **Look at mistakes differently.** See them as lessons, not failures. It took me a while, but eventually I saw my call sign error as the lesson I needed to learn. It was what I needed to know early in my career, and it helped me become the fighter pilot I needed to be.

3. **Look for the silver linings.** When you create a new perspective, you can open up new opportunities. Train yourself to always look for the good inside the not-so-great and you'll likely discover new avenues for growth. At the very least, you'll feel less defeated.

4. **Enable serendipity spotting.** Keep your eye out for new information, opportunities, and ideas that can benefit you, such as a neighbor in a race bib who, it turns out, joined a new running club you didn't know about even though you've been searching for one.

5. **Leverage technology.** Thanks to social networks, we can reach a wide range of people with connections and ideas that can help us. LinkedIn is a great place to use technology to your advantage. Whenever a colleague starts a job or earns an accolade, check in with your congratulations. You never know when they might need to hire someone like you.

Several inventions are the outcome of serendipity or of someone recognizing the serendipitous trigger and tenaciously connecting the dots. Percy Spencer was an engineer for Raytheon working on radar equipment for the military during World War II. He stopped for a moment in front of a magnetron, which generates high-frequency radio waves. Suddenly, the chocolate bar in his pocket began melting. So he stuck some popcorn kernels near the magnetron and, you guessed it, he got popcorn. Then he put an egg in a pot near the machine and wound up with a cooked egg.

He could have left it as some fun parlor trick, but he and other company engineers instead went on to develop the first microwave oven.[12] His discovery-turned-invention supports this eye-opening stat: 50 percent of success in many areas is due to an "unexplained variance." If he hadn't stuck a candy bar in his pocket, we might not have microwave ovens to reheat dinner. But it took more than that moment for the outcome to come to be. It took his curiosity and determination, and it took financial and development support by Raytheon.

The idea of serendipity leading to scientific breakthroughs is so prevalent, the European Research Council reportedly gave social scientist Ohid Yaqub a nice, signed grant of nearly $2 million in U.S. dollars to study serendipity in science.[13] He found four types of serendipity, each named after a scientist:

1. **Walpolian:** Finding something by chance without looking for it, as in the discovery of penicillin by leaving mold in a petri dish.
2. **Mertonian:** Finding something by chance while looking for something else, as in when a pharmacist looking for a cure for headaches accidentally invented Coca-Cola.
3. **Bushian:** Finding something by pursuing curiosity, as in Spencer's melting chocolate bar (then popcorn, then cooked egg).
4. **Stephanian:** Finding something by interacting with others, as in the Oxford–AstraZeneca Covid-19 vaccine, created by researchers in different fields.[14]

It's not enough to have the serendipitous moment. You have to have courage to take advantage of it. University of Hertfordshire professor Richard Wiseman, author of *The Luck Factor*, created "luck school" to help increase the amount of luck people experience.

First, he used questionnaires to measure the self-reported levels of luck. Participants assigned a measurement from a scale of one to seven, ranging from "doesn't describe me at all" to "describes me very well." Then they were given descriptions of people who are "lucky," defined as chance events tending to work out consistently in their favor, and "unlucky," where chance events tend to work out consistently against them. Each participant was then classified based on their scores as either lucky, unlucky, or neutral. Then he described his four principles of luck: create and notice chance opportunities, listen to intuition, use positive expectations, and maintain a resilient attitude. Then the participants spent a month on various exercises and returned to describe the outcome.

He found that 80 percent of the participants reported feeling happier and luckier. "Without realizing it, lucky people tend to use various techniques to create chance opportunities that surround them."[15]

You don't have to be a scientist or researcher to create luck. Think about how you can apply these principles to your own life to make your own luck factor.

PIVOT!

One way to not only survive sudden changes but to take advantage of them is to learn how to pivot, when you change your plans without abandoning your overall vision. The pandemic was ripe for pivots as people had to figure out other ways to live and work during lockdowns. You've no doubt heard stories of distilleries that used the alcohol that was no longer going to bars, which were closed, to produce hand sanitizer. Or the deli owner who shifted to a grocery business that's still growing today.

We, too, had to pivot on the Thunderbirds. Suddenly, we became an air show team with no air shows, but we still had to fly to stay proficient. So we came up with an idea for a pivot: Operation America Strong, when the Navy's Blue Angels and the Air Force's Thunderbirds conducted a series of multicity flyovers during the spring of 2020 to salute health care workers and other first responders.

These were some of the hardest flights during my time with the Thunderbirds, seven-plus hours each with a ton of coordination, multiple air refuelings, and an acceptance of a higher risk level than the average air show, where all the maneuvers take place inside a TFR, or temporary flight restriction. That means no one else is airborne in that area, so we don't have to worry about deconflicting in the middle of a precisely timed routine.

Outside of a TFR, it's a bit more of the Wild West. During America Strong, we spent much longer in formation, operating two different aircraft, our F-16s and the Blue Angels' F-18s, in close proximity. Each had different ideal airspeeds and maneuverability. Each team also has its own procedures.

During our first long America Strong flight, up the front range in Colorado, the clouds were low and it was snowing, creating a disconcerting lack of depth perception as I tried to focus on the wing tip of the aircraft next to me. I worried that I couldn't keep the focus needed to continue to fly in formation.

In later flights, we would deal with air traffic controllers who weren't used to juggling two six-ship formations traveling at more than three hundred miles per hour, close calls with birds, and a near midair collision with a helicopter. The mission was rewarding, but the risk was high.

The America Strong flyovers were something that the country needed during the height of the pandemic, and for some people they were the only time they got to see us perform. Now when I'm on the road or in client meetings, people often tell me they found the Thunderbirds and eventually, my Instagram or LinkedIn pages, thanks to seeing us fly over their part of the country. Our pivot during the pandemic even led me to be hired as a speaker.

I've pivoted multiple times, and it took flexibility and the ability to bridge situations and circumstances to make it successful. I'd thought I was supposed to become an FBI agent, so I majored in criminal justice. I enrolled in Air Force ROTC as a way to pay for college, figuring I'd use the experience as a pathway to the FBI or maybe the CIA. Then, I figured I'd live happily ever after chasing bad guys.

Despite my well thought-out plan, that was not what happened. I witnessed those F-15s blasting off for a training mission in full afterburner and something in me shifted. So did my goals. After college and after I received my commission in the U.S. Air Force, I was selected for a pilot slot and, against the odds, was assigned to fly F-16s…yada yada…I became a member of the elite Air Force Thunderbirds.

The "yada" in this case was taking a lot of pivots and seeing opportunities in the serendipity. It was my adaptability to unexpected opportunities, including raising my hand and declaring that I wanted to become a fighter pilot, volunteering to train pilots in Poland where women had never flown F-16s, let alone trained pilots to fly them, and taking time off to hike in Nepal.

This isn't some superpower of mine. You can learn to position yourself to pivot toward luck as well. You can make smart changes, first by embracing change with curiosity and optimism. Change is where growth happens, so welcome it, and then be willing to pivot when an unexpected opportunity emerges. It may not be something you planned for, and that's okay, because adaptability helps us find our true purpose. That's why you should set goals, but don't become married to them.

Remember, success and fulfillment can come from the most unexpected places. If we learn to adapt, pivot, and recognize these moments, life becomes more enjoyable and opens up endless opportunities from often seemingly serendipitous moments.

THE FLIPSIDE

Scriptside:

- Assuming that luck happens only to "lucky" people who are born that way.
- When luck comes your way, shrugging and chalking it up as a rarity.
- Forming your opinion and sticking to it.
- Surrounding yourself with people who don't want to grow.
- Looking at pivots as necessary evils.

Flipside:

- Using the five dimensions of curiosity to bring luck your way.
- Remembering that smart luck is all around you if you just look for it.
- Connecting the dots to make serendipity work for you.
- Creating pinch-me moments with a portfolio view of friends and colleagues.
- Pivoting!

CHAPTER TEN
Avoid Get Home–itis

We've just finished one of our last air shows of the season and I want to get home so bad, but my jet's hydraulic gauge is fluctuating. It's staying within the operational limits, so no warning lights are coming on—yet. It's a long flight home from the eastern United States, some four or five hours long, and we're a little over halfway back. I keep watching the gauge but keep my concern to myself. I'm not breaking any rules because the levels are within the limits, but it's a bit of a gray area because things aren't quite right. This is a situation where you rely on your experience and judgment, keeping a close eye on the evolving environment around you.

I'm concerned about these potential jet issues, but I can feel the emotional weight of the entire team on my shoulders: They want to get home. The Thunderbirds average 240 days on the road per season. Combine this with the energy demands of travel, flying multiple shows and practices, and engaging as public figures, and it's no wonder we always look forward to coming home. When one of us stops to get an aircraft fixed, our wingman has to stop, too.

THE FLIPSIDE

As I continue to watch the needle wiggle back and forth, I map out the most preferable divert options between here and home. Should we divert to Colorado or somewhere in New Mexico, or keep flying back to Las Vegas?

A half hour later, somewhere over the Grand Canyon, the gauge starts swinging below the limits and suddenly, the warning lights come on. Now I have an impending hydraulic failure on my hands. It's the system that controls my normal landing gear and braking—for a jet that touches down at over 150 miles per hour. This has become a big deal, and it's very clear I need to tell my flight-mates that I'm having a problem. I'm in a heightened state of focus, but I'm not panicking. This is one of those emergencies that we regularly train for, and the jet isn't going to suddenly fall out of the sky. I have time to run through the checklist, collaborate with my team, and pick the best course of action.

Our home base, Nellis Air Force Base, is a good option now, as we're not that far out. I declare an emergency, coordinate with the control tower, and lower my landing gear with an emergency backup system used in case of a hydraulic failure. Even though I have time to run through my options, this malfunction does mean there is the potential of issues during landing, like that the landing gear could collapse on touchdown. If your brakes aren't working well, you also may not be able to stop your jet before it skids toward the end of the runway, and the backup braking system is good only for a short, steady application. I'll have to stop on the runway and stay there until a special team from the base comes out and tows my jet back to the hangar. That means I will shut down that runway for a decent amount of time, and the only other runway at Nellis is closed for maintenance. Other local training flights are soon diverted to Creech Air Force Base about forty-five miles north of town, no doubt aggravating the pilots who also just want to get home.

You could argue that I should have landed immediately the moment the gauge began fluctuating, but a nonmilitary airfield might not have the necessary equipment to stop a malfunctioning F-16. It wouldn't have our maintenance support or a nice, long runway or, let's face it, crash support. I was in the gray area where I had to make decisions, so I drew the line for myself: As soon as this hydraulic gauge drops outside of the operational limits and triggers the warning light, I'd shift from considering options to taking action.

I could feel a case of what pilots call Get Home–itis creeping in, a tide of emotions overwhelming the hard facts. It's like that summit fever I mentioned, when mountaineers throw their safety criteria to the wind to push to the top despite new or unexpected risks. Get Home–itis is when our desire to get somewhere outweighs our logic, decision-making, and risk assessment.

We'd all suffered from it before. Once, in a previous air show season, a convective weather system with turbulence, icing, and low ceilings had wedged itself between Las Vegas and a show the Thunderbirds were headlining in Northern Texas. We were often the big draw that show sites and host cities relied on for their big investment to be a success, so it was important for us to get there.

If we pushed the flight twenty-four hours, the weather would be better, but we'd miss our practice and possibly lose our tanker support for air refueling, forcing us to hop from airport to airport, and then we'd risk facing a maintenance issue without the support to fix it. It also meant we wouldn't have a chance to check out the terrain, the references, and the obstacles before the show. And, let's be honest, shifting a day later is a pain in the butt, so no wonder there's always a lot of pressure to get there on time. How much weight do you put on that?

Sitting here at one G and zero knots, it's easy to say the solution is obvious: Delay the departure for the air show by twenty-four hours. It would be hard to justify flying into a thunderstorm just for an air show. However, that is a massive oversimplification. When we are in a Get Home–itis scenario, we are at our most vulnerable. Emotions have infiltrated the equation, and it's tough to explain how much weight this variable can have, how much it colors our decisions. So how do we avoid falling into a scenario where we make a riskier decision than we usually would?

GO/NO-GO

In the cockpit, we determine what we call Go/No-go scenarios before a mission, depending on the acceptable level of risk. We do this before we take off because that's when we're acting as rationally as possible. That way, our decision-making isn't swayed by the heat of the moment when things go sideways.

In a combat scenario, if the circumstances change, we've already set the risk criteria in advance. So if we're suddenly outnumbered or it appears we're now on a suicide mission, we've already determined we're either turning around or not taking off in the first place. The risk isn't worth the reward.

We didn't fly a single flight where we failed to talk about contingency plans beforehand. We'd plan out what to do if we had four good guys and one had a maintenance issue and had to turn around or if one found a malfunction before takeoff leading to a ground abort. We'd determine the acceptable level of risk for a mission depending on the objectives and the urgency. We'd rate the risk of each flight based on the weather, the proficiency of the pilot, and the stress level of the specific mission. And, because it was the military, we assigned numbers to it all. If we had a training flight and there was, say, a tornado warning, the pilot was sleep-deprived thanks to a new baby at home, and it was a complex mission, that number would skyrocket based on the point value and the decision would be obvious. If those numbers hit a certain level, the mission required approval by those higher in command.

Outside of the military, I've had Go/No-go turn me around on a mountain—three times. Not long ago, I climbed Mt. Hood. Snow had dumped on the mountain for two days straight, but on summit day, the sky was crystal clear. We started climbing in the middle of the night, and I was a little worried about my knee, which had been bothering me. Still does. But that night, my knee held up, and I felt really good.

My group reached the technical part of the mountain just below the summit where we needed an ice ax and crampons to cross steep terrain. We were the second group in line, watching the first group break trail, when we all saw little sheets of compacted snow breaking off the surface and sliding down the mountain, indicating a potential avalanche risk. With several groups behind us and one in front of us lined up like a trail of ants, an avalanche would have taken out a dozen people, depositing some of us into the fumarole, a hole emitting toxic volcanic gasses, below us. In fact, a snowboarder had landed inside it not long before our climb and had to be rescued.

With one call from the person at the front of the group that the snow conditions didn't look good, we all turned around. We'd hit our No-go, just shy of the summit. I was disappointed, sure, but it didn't change our Go/No-go criteria, which had already been set before we even started climbing.

It wasn't my first mountain No-go. It also happened to me on Mt. Baker and then on Mt. Rainier. Every single time we'd get pretty dang close to the summit, and every single time, we had to turn around, once for snow conditions, once for weather, and once for a climbing-mate who'd fallen ill.

When I tell people I've climbed these mountains, I don't preface it with "but we got turned around an hour below the summit." If we get into the details, of course I tell them, but to me, I'd achieved my goals because of how I'd defined a win. Would summiting have been a win? Of course! But I had other wins, too, like learning how to use an ice ax or use crampons on icy slopes.

Our decisions relied on our skills and training and our level of trust in our leader and in each other. We looked at risks objectively in advance, not when the summit was tantalizingly close, helping us set our expectations before we even put on our crampons. We also understood the seriousness of the wrong decisions. Finally, once we created our No-gos, we had no reason to question our decision to turn the heck around.

Our Go/No-gos are important because they help us avoid a major contributor to Get Home–itis, the sunk cost fallacy. It's when you choose to continue a plan or an event based on past investments of resources, like time or money, despite evidence that the venture no longer makes sense, which is a form of commitment bias. It can be something as common as finishing a bad meal in an expensive restaurant or plodding through a boring movie because you paid to rent it. But it can be something more consequential, like staying in a relationship that stopped working years ago or pursuing a career you've come to dislike simply because you're still paying off the tuition for the major that led you to this industry. Even though the circumstances have changed, you remain committed just because you said you were going to do it. You've become irrational, letting emotions color your decision-making.

SETTING EXPECTATIONS BY MAKING RATIONAL DECISIONS

Making rational decisions ahead of time not only makes sense in the Air Force or on a mountain, it's also a recipe for a more peaceful life.

For instance, what if your daughter's soccer team has a bad day and gets trounced in a game and she announces she wants to quit the sport? Maybe you tell her she can quit, but not today. She's got to make that decision on a good day, like after a big win or during the team end-of-season party. You're not saying, "We are not quitters," because sometimes, quitting is a perfectly fine decision. But you're not letting charged emotions influence the decision, either. It's easier to be rational when you're not in the heat of the moment.

One way to start making rational decisions is to set small goals, even for smaller things. For example, if you sign up for a three-hour course on personal development, and one hour in, you're feeling like it's a waste of your time, how do you know if you should leave or log off? If you had decided beforehand that you want to learn a specific lesson or understand a certain concept and the class is not as advertised, then it doesn't achieve your goal. It's a No-go. Leave the meeting and do something that serves you better with that time.

But wait! What about your investment? The money charged on

your credit card and the time spent? You even combed your hair for the Zoom! Whether it's something short, like a class, or longer, like a career, I ask myself what I could have accomplished toward my goal in that time. I challenge myself to think, *Six months from now, I could have six months of progress in that new job/training/relationship/endeavor/goal*, instead of continuing to slog through something that isn't working despite my best efforts.

For fighter pilots, much of our decision-making comes in the preparation phase, the mission-planning part when we consider all our options before a flight. If there's bad weather expected, we discuss whether we can fly above it or divert around it. We determine if we can take off earlier or later to miss it, or if we can land at various bases to refuel instead of using the midair tanker. We decide ahead of time that if we have a maintenance issue, we will find the best place to land, with a wingman joining us, no question about it.

You can do the same thing with personal decisions. For example, if you've reached a dead end at your job and you're no longer as passionate about or invested in it, ask yourself if there's a different role at the same company you can accept. Can you take a sabbatical? Can you stick it out longer or work part-time? Is there a job you can apply for elsewhere? I asked myself all these things before making the decision to leave the Air Force.

Quitters do, in fact, win and winners quit, but that was hard for people to accept when I left the security blanket of the military pension behind for my career in speaking and writing. I knew I wanted to leave active duty. My body was beat up, and flying fighter jets part-time wasn't an option I wanted to pursue, because I only wanted to do it if I was 100 percent present, focused, and proficient. I considered the National Guard or the Air Force Reserve, and I ended up working part-time for Air Force Recruiting via the Reserves for a while.

I soon discovered that I couldn't give the role the attention it required. I was building my business, booking out speaking gigs months in advance that would ultimately conflict with reserve duties. I couldn't show up in the way the position deserved, and I realized it would be better filled by someone who could. So I quit, fourteen years into the twenty required to earn retirement benefits, thereby sealing my fate: no military pension.

At times, I felt like I was letting people down. And I was told again and again how people didn't understand how I could be foolish enough to walk away so close to retirement. Internet trolls were the worst because they called me a quitter, among other things. But I wanted to go out on a high note instead of letting my career peter out by half-assing it with too many things on my plate. I wanted to do something that I felt excited and passionate about, and I wanted my back pain to stop being exacerbated by nine times the force of gravity. Besides, as soon as I let that part of my life go, the rest took off. Pun totally intended.

I avoided the sunk cost fallacy by making rational decisions based on my current goals rather than deciding the only thing that mattered was my old goal of reaching twenty years in the Air Force and full retirement benefits. Much like trying to reach the summit of a mountain, there were many benefits to appreciating the journey along the way, and it had value independent of reaching what many viewed as the benchmark of a successful military career. This viewpoint gave me permission to consider what I really wanted and to go for it.

Changes happen all the time—new jobs, new relationships, new homes, new opportunities, new challenges—and you'll need to make plenty of decisions. In fact, it's been reported that we make some thirty-five thousand conscious decisions per day.[1] Some of our decisions are automatic, like which way to turn off our street to get to the supermarket or whether we should plant a kiss on the dog's snout. These are based on habits, emotions, and instincts. But others are more deliberate decisions based on logic and reasoning. Still, even deliberate decisions can be clouded by our biases about the types of data we're considering.

One way to make rational decisions is to understand the different types of data and your biases toward them. *Harvard Business Review* identified three types of data to consider:

1. **Contextual data:** Anything external that can affect the situation, such as weather, news, the economy, public sentiment, social media, etc.
2. **Patterned data:** Information that follows a structure or form, when we might assume that past events predict future occurrences.
3. **Salient data:** Relevant information given the context, which can lead us to give too much weight to new or notable information.[2]

For instance, when two upcoming speaking events with big clients that I was excited to work with were canceled in one week, I had a moment of panic. Was economic uncertainty causing companies to cut spending on events? Was my message no longer resonating? Did I need to completely change up my marketing strategy? No. The cancellations were unrelated to one another and new events continued to book despite the weight I was putting on the two I'd lost. For a moment, I failed to make rational decisions, letting my feelings about new data skew my perception and, therefore, my decision-making.

WHAT'S YOUR WHY?

Not all decisions you make are about practical situations. Sometimes, you have to make rational decisions about emotionally charged circumstances, but you can still apply some logic to sort through what matters to you. You do that by filtering all decisions through your "why," the motivating factor behind all you do. Call it a motto or a raison d'être, it's what gets you out of bed in the morning.

Top Gun's Maverick might have declared he had "a need for speed," and that's fun Hollywood dialogue, but my "why" in the fighter jet was the same why I have in my life: to live authentically while making a positive impact for others, and to have a blast doing it. That has been true from the day I set a goal to become a fighter pilot, and it remains true in my motivation for pivoting out of the military and for writing this book. How that positive impact for others is created has evolved over time, but it has always been part of my why.

I've asked myself why I'd joined the military, why I wanted to fly, and what was most important to me. The answer: I had wanted to see the world, be part of an exceptional team, and serve my country. I wanted to do something that felt fulfilling and exciting. By the end of my flying career, I had done all of those things and more.

I learned so much, so fast, as a fighter pilot, and a lot of what I learned applies to everyday life. You may not need to know how to engage in a dogfight up in the air, but you can use a lot of the mental skills and coping hacks that fighter pilots develop. My why centers around teaching others what I've learned, all the while being my true self and having fun.

Along the way, I realized that I had found even more fulfillment in directly inspiring other people, and I often discovered it while meeting people one-on-one along airfield fences after a show. During my first season with the Thunderbirds, we performed at a show in the Midwest, landed, and walked to a designated spot along the fence to meet and greet the crowd and sign autographs. I had begun to realize that my time signing autographs would continue long after my peers had finished. Not because I did anything different in the air, but because mine was full of families eager for their daughters to meet someone who looked like them accomplishing something as impressive as flying F-16s overhead.

As one mom pulled her six-year-old daughter toward the fence, I recognized that the girl, eyes downcast and hesitant, was embarrassed and shy. When they finally reached me, the mom said, "Look, Anna, that lady flew the jets you just watched." Anna slowly looked up at me, connecting the dots between what she had witnessed during the air show and a person that she could relate to. A light turned on in her eyes. Over the next few minutes, previously shy Anna bombarded me with excited questions about flying upside down and spinning my jet around in the air. By the time the autograph session ended, her mom had to drag her away.

I had become a planter of seeds of possibility for others, and when that was no longer something I could do most effectively in the military, it was clear it was time for a pivot to speaking and writing. The moment the new position failed to align with these goals, I abandoned my plans. It was clear that it was a No-go because it no longer fit my why.

How do you determine your why? Entire books and courses have been written to help people find their why, and Simon Sinek's *Start with Why* is perhaps the most well-known. His TEDx, "How Great Leaders Inspire Action," introduces the idea as it relates to business and leadership.[3] He said that most people who work for companies know what they do and how, but not all understand why.

If you ask the average person why they work, they'll reply, "To pay my bills." But that's the what and how, not the why. And this doesn't just have to be about work, either. There's a why behind your decisions, big and small, and it goes beyond the practical. You just have to know how to identify it. Once you do, you'll make decisions that fuel it.

There was something deeply visceral about seeing those jets' afterburners when I was standing on the tarmac during my Air Force ROTC training in Florida. When I felt the rumble of those engines and something inside me declared, *I want to do that*, I could have chalked it up to a pipe dream. Instead, I honored what my gut was telling me, paid attention to that spark of inspiration, and **I followed my gut**.

I couldn't yet put my why into words like I can now, but I knew that declaring that I wanted to fly fighter jets was the first step toward achieving it. Having the courage to **say it out loud** created accountability for me to take actions that supported following my gut and gave others the chance to help me along the way because they knew what I wanted. It set off the string of actions that ended with me in the cockpit of an F-16.

Then I had an opportunity that led me to give my why weight. When I was stationed in Japan, I had the chance to fly a spare F-16 as a safety observer for the Pacific Air Forces aerial demonstration team. Based in Misawa, the team performs in air shows throughout the Indo-Pacific region to promote goodwill with people in area nations. As the safety observer, you don't get to perform like the demo pilot. You're just bringing another jet in case they need one to places like Malaysia, Australia, and India. But you do get to engage with the crowd, and often, they'd never seen a female fighter pilot before, and my presence was obviously impactful to many of the women and girls who attended our shows. I saw the looks on their faces and answered their eager questions.

This was me **giving my why weight**. I soon recognized that whenever I did something that felt authentic and aligned with my motivations, I could hone in on my why until, eventually, it became very obvious.

Later, I'd use this well-defined why to guide my decisions while allowing it to evolve through new experiences, priorities, and moments of inspiration. As we grow, we evolve, but I'm still positively impacting people, being myself, and having fun. I'm just not doing it from the air at the moment.

MIX IN YOUR CORE VALUES

The foundation for my why is made up of my core values. Companies often make a big deal of theirs, and it helps us to understand what they stand for. Whether it's Google's promise to commit to "significantly improving the lives of as many people as possible," Netflix's aspiration to "entertain the world," or the Cheesecake Factory's promise to be "relentlessly focused on hospitality," their values reflect who they are.

The military, too, promotes core values that inform their behavior and decisions. Perhaps you've heard of the U.S. Marines' "Semper Fi," which is short for "semper fidelis," or "always faithful." The Air Force's core values are: integrity first, service before self, and excellence in all we do.[4] This makes clear what our branch's principles are so we can align ourselves to its why.

The Thunderbirds, which are part of the Air Force's Fifty-Seventh Wing, are considered "America's Ambassadors in Blue" and are often referred to as "America's Team."[5] Their mission is to recruit, retain, and inspire, and they accomplish that under the umbrella of the Air Force's core values. These values aligned with my personal values, especially to inspire, and were one of the things that drew me toward the team.

Individuals also live by core values, and when you know what your values are, you can align your actions with them to live more authentically. Some examples are: adventure, autonomy, community, kindness, and wonder. If you haven't labeled your own, it can be helpful to see some examples, and a quick internet search of common core values will give you infinitely more choices until you find the ones that feel right.

If you're not sure which values to choose, think about someone who embodies all that you'd like to be. It can be someone famous or someone in your social circles or your family, someone living or dead, someone who has achieved what you'd like to achieve or who behaves the way you'd like to be. There's a good chance it might be someone from your Dream Team because they likely align with some values you admire.

You can even level up from relying on your Dream Team to relying on an alter ego for yourself, like Beyoncé did when she pretended to be a character she made up named "Sasha Fierce." She explained that Sasha was who she became to transform from her shy, quiet self to her, well, fierce onstage persona.

Author Todd Herman, too, believes that an alter ego can be a means to success. He teaches high-level performers, such as professional athletes, to choose secret identities that encompass all the qualities they wished they had, just like we did when we were kids. Whether you create your own character or choose one that already exists, your alter ego should embody the core values that matter to you. So if justice, equality, and strength are at the top of your list, perhaps you might choose Wonder Woman as your alter ego. Are you big on curiosity, knowledge, and innovation? Albert Einstein might be your secret identity. You can even be Sasha Fierce now that Beyoncé is no longer using her: "I don't need Sasha Fierce anymore because I've grown and now I'm able to merge the two."[6]

We can't go it alone all the time, and sometimes it's nice to have an alter ego or a Dream Team member to remind you what your Go/No-gos are. Look, Get Home–itis is easy to catch, especially when you've been working hard and you need a break. But that's exactly when you're more likely to skirt your values for an easy reward or fall prey to the sunk cost fallacy. I mean, you've already invested all this blood, sweat, and tears into your goal, and you never set the No-gos that force you to recognize when you need to make an emergency landing. No wonder you just keep rushing toward something that no longer makes sense. It feels easier than considering other options. But if you set your Go/No-gos so that you can make rational decisions, and remember your why, you'll avoid Get Home–itis.

In time, you'll be ready to capitalize on moments of inspiration, your version of the jets taking off or the reward of empowering those kids at an air show. Let whatever that is inform your why so that you have a guide to come back to even when stress is high and you just want to get home.

Scriptside:

- Waiting to make important decisions until you're in the heat of the moment.
- Failing to identify the "why" behind your decisions and goals.
- Failing to sort through what matters to you most.
- Mindlessly continuing toward a goal despite changed circumstances.

Flipside:

- Determining your Go/No-go.
- Defining your why.
- Choosing your core values.
- Being willing to pivot when your goals no longer align.

CHAPTER ELEVEN
Check Your Six

It was late summer 2021, and the Taliban was coming for anyone who had worked with the United States as our troops prepared to withdraw from Afghanistan. They were coming for one of my pilot training classmates. They were coming for Taj.*

Suddenly, I was added to a Facebook chat of two dozen pilots who had trained with Taj, an exchange pilot from Afghanistan, a U.S. ally. He had become friends with all of us as we navigated the yearlong pilot training course, but since then, we'd mostly lost touch with him. Though he'd left his home country for Canada, he had recently flown back with his new wife to visit his family, including his father, a high-level Afghan National Army officer. In the Taliban's eyes, this placed a target on his father's back and on his entire family.

Taj needed our help. He had fourteen family members, including his mom—a diabetic—some young women, and several children, to hide from the Taliban, who were once again taking over

---

* Alias.

the country two decades after the United States had ousted them from power. Taj and his family had left their house to go into hiding, and they watched on a security camera feed as the Taliban raided their home.

Taj was desperate, so of course we'd do everything we could. That's what wingmen do. Some Afghans were able to get on lists to escape as refugees on American C-17 cargo airplanes leaving from Kabul, while others were seen hanging off the aircraft during takeoff, risking almost certain death rather than live under Taliban rule.[1]

We all started emailing and calling our connections to try to get the family out of Afghanistan. I started with the Thunderbirds, where I was in my third season, and eventually turned to my social media audience. One of my classmates was working his NATO channels, and others were cold emailing retired generals, bypassing the rule for military communications that followed a strict chain of command. But these were not ordinary circumstances, and each hour that Taj and his family remained in Afghanistan posed a significant threat to their lives.

My husband reached out to a connection with the FBI, and within twenty-four hours, he received a call from an unknown number. It was an intelligence agent from an ally country, and he had a very generic name (ahem, alias), like Tom Smith. The next thing I knew, I was collecting email copies of the family's fourteen passports and funneling them to "Tom."

Meanwhile, my social media plea found its way to someone with contacts at the White House, and long story short, twelve of Taj's family members, including his parents, ended up outside one of the gates at Kabul's airport. One of the few other ways out of the country, Bagram Airfield, once a bustling city under American control, had already fallen to the Taliban. That left Kabul as the only escape route by air.

His mom was out of insulin, so we were trying to get her medicine and supplies, as the family spent the night sleeping outside the gate with hordes of other people. Just one day after they got pulled into the airport, a suicide bomber killed 170 Afghans and thirteen U.S. service members outside that very airport.[2]

Taj and his wife were racing to meet their family when they got stopped at a Taliban checkpoint where their phones were confiscated and smashed. They were warned: Come here again and we'll kill you.

For twenty-four hours after, Taj was completely out of touch with us, and we were all frantically messaging each other for news about our friend. We began to think the worst had happened.

His family, meanwhile, made it out on a C-17 and eventually arrived at one of the main refugee holding locations, Fort McCoy in Wisconsin. I reached out to the congressional representatives in my home state to try to get information about their immigration status and what the next few months would look like. We rallied to get them supplies, Venmo-ing money to a distant contact who'd seen my social media plea and had the necessary access to get on base with a shopping cart full of clothes and baby formula.

Taj, meanwhile, was still MIA. Then we got an email. Taj and his wife were able to get online but were no closer to having a safe way out of Kabul. His wife had been so freaked out by the Taliban checkpoint interaction, they decided instead to make a run for the border. Only, Uzbekistan was turning people away, so we told them to stay put while we coordinated a military team to get them to the airport. Nevertheless, they turned down the offer and left for the border with no phone or internet.

For another twenty-four hours, we heard nothing from them. Just silence. Some two dozen American service members, most of whom had not had any contact with Taj for a decade, held our breath as we hoped for good news yet braced ourselves for bad news. I struggled to focus on my work, anxiously refreshing my messages again and again in between briefs, air shows, and PR events. Everything outside of Taj's escape from the Taliban seemed insignificant.

The next time we heard from Taj, he was standing in front of a sign in Abu Dhabi in the United Arab Emirates. His message: "My wife and I made it. Safe and alive."

Eventually, they were able to cross the border, buy tickets on a commercial airline, and fly to a safe country, thanks to their determination and a network of wingmen who wouldn't give up.

YOUR COMMUNITY OF WINGMEN

Remember: Fighter pilots never fly a mission alone. We always have a wingman, another jet flying with us. It's not just for the firepower of multiple aircraft, but also because there's a lot going on in a single-seat fighter jet, and it's helpful to have two minds solving any problems. In the air, we have many decisions to make beyond flying from point A to point B, and we need backup in case the unexpected happens.

So we watch each other's six, that spot directly behind the jet. That's so you don't have to twist around to see the blind spot—and possibly, the bad guy—behind you. When I had the hydraulic failure on our way back to Nellis Air Force Base, the Thunderbirds sent my wingman, Flack, to fly alongside me for support. He read the checklist for the technical issue and called ahead to the air traffic control tower to warn them of the emergency.

Outside the cockpit, wingmen are the people you know you can call when the shit hits the fan, the top names on your "recent calls" list, and the first people you text when things go bad or well. But they can also be people you haven't talked to or seen in a while, like the friends and colleagues you message through Facebook when your family is trying to escape Afghanistan, for example. Sometimes wingmen just show up, like the neighbor down the street who's had the same diagnosis as the one you just got, or everyone in your new support group, or your fellow recruits and first-timers.

Early in my career, I resisted letting my wingmen help me because admitting that I needed help felt like weakness. This slowed my career progress and left me struggling all on my own. It was also a very lonely way to live. It would take me a long time to learn just how important community is to individual success, because I long believed that I should just show up, do my thing and, pardon the pun, fly under the radar.

I didn't fit into the culture of a fighter squadron no matter how much I tried to belong, walking a difficult line between proving that I deserved to be there and never stepping beyond slightly above average. I didn't want to stand out too much or my peers would resent me, so I kept to myself instead of building my community. This was a big mistake, creating obstacles where there didn't need to be any, starting with loneliness.

The irony is that when it comes to loneliness, we are not alone. It's widespread, and as a nation, we've never quite understood the negative effects of loneliness as much as we did during the pandemic. In a 2021 survey by the Harvard Graduate School of Education, more than a third of respondents reported feeling lonely "frequently" or "almost all the time or all the time" during the prior four weeks.³ An alarming 61 percent of adults ages eighteen to twenty-five reported high degrees of loneliness.⁴ The survey showed that no one was spared—no race, ethnicity, gender, education level, income, religion, or location.

By 2023, U.S. surgeon general Dr. Vivek Murthy issued an advisory on the country's "epidemic of loneliness and isolation," linking them to physical health consequences, such as an increased risk of heart disease, stroke, and premature death.⁵ You'd think we'd have been scrambling to get together after a few years of being trapped at home, making sourdough starter, but it's not so. The U.S. Census Bureau reported in 2023 that Americans had actually spent less time with friends and more time alone since before the pandemic, with loneliness increasing "linearly" since the 1970s, reported *PBS NewsHour*.⁶

These days, we entertain ourselves alone with Candy Crush or Hulu on our individual devices, not even gathering together to watch TV. We sit alone in our home offices on Zoom calls and stare at our devices in restaurants, on trains, and at crosswalks instead of interacting with one another. When we want to learn something, we don't call up a knowing friend or a smart uncle; we google it. And my generation doesn't make phone calls. If you call us, we'll ignore it and then shoot you a text that says, "What's up?"

No wonder we're lonely. But community is something we all need and can have, if we build it. You may not have people who immediately come to mind when you think of wingmen. Don't think that makes you weird or different, because the stats bear that out. Fewer adults report having a best friend compared to thirty years ago,⁷ and Americans have fewer friends overall, with half reporting three or fewer friends compared to about one-quarter back in 1990.⁸ Social media can keep us in touch with our second cousins and our high school boyfriend's dad, but it doesn't seem to cultivate deep friendships. We keep in touch with more people, and somehow, we're not as connected.

As a result, we have to deliberately and methodically build our own communities of wingmen. But first, you need to find your people. I've done that through my interests and hobbies because if I like to do something that other people like to do, too, we have an automatic bond, an initial common ground that we can build from. It's a matter of finding people who'll pull you back in the boat when you need it. I should know.

On a weeklong whitewater rafting trip in Deschutes, Oregon, I met a bunch of other military veterans of different branches, ages, genders, and races. We rafted by day and camped on the riverbank at night. They weren't all fighter pilots like me. They were Marines, sailors, all sorts of veterans, but we all shared a love of outdoor activities.

One day, we were rafting on class-four rapids when we hit a hole that caused the raft to bend downward and spring back up, and all of a sudden I was underwater. It happened so fast, I had no chance to even grab on tighter or dig my foot in deeper to try to stay in the raft. It was just paddle, paddle, paddle...underwater, looking up at bubbles. A few seconds later, I popped to the surface. Immediately, one of my fellow veterans flipped his paddle around and shoved the handle toward me. I grabbed it, and moments later I was hauled back into the boat.

What do you know? My wingman was former security for Marine One, the president's helicopter. This is a great example of how you never know who'll pull you back into the boat. We weren't close and we haven't kept in touch since, but we had common ground and worked well as a team. We had mutual respect and common interests, and when one of us really needed some help, there was no hesitation to offer it or accept it.

Most important, we had the two main ingredients of building community: curiosity and vulnerability. He had curiosity, as in "Where the hell did Mace just go?" And I had the vulnerability to accept help. He cared for my well-being and I accepted the help, which was for me a big change from my early days as a pilot when I might have ignored the paddle in my face and attempted to swim for the raft on my own. I had a hard time being vulnerable, even around people in the very same situation I was in.

Dangerous circumstances aren't the only times community is important. When I was a brand-new lieutenant in the Air Force, I really wanted some female friends in the mostly male community I'd joined, but instinctually I saw other female fighter pilots as competition. I'd see a woman from another squadron at an exercise and immediately start sizing her up: *Is she farther along in her training than I am? How does she fit in with her squadron? Do they respect her? Is she prettier than I am?* All that primal, competitive stuff rushed in. It wasn't about being the best fighter pilot. It was about being the best *female* fighter pilot. We'd been inundated with the idea that there was one spot at the table for the token woman, and it fueled this competition.

Then we'd start talking and I'd find out she felt the same way about the atmosphere in the squadrons and had been through some of the same things. "Oh my gosh! You deal with that, too?" Suddenly, we had a sisterhood rooted in a shared experience and a deep understanding that no one else on my team had endured. It was like instantly feeling seen. After being proven wrong time and time again when I made the effort to actually get to know these women, I eventually made a conscious decision to recognize my initial, competitive, gut reaction and override it, telling myself, *No. Go introduce yourself.* The vast majority of the time, I'd make a new friend.

CURIOSITY + VULNERABILITY = COMMUNITY

When you are curious about people and you're willing to share who you really are deep down, people reciprocate, and then you can begin to build community. But this is becoming increasingly difficult in a cancel culture that doesn't value deep listening and understanding. It's not without consequences. In fact, several long-term studies found that a lack of curiosity is reportedly lopping years off our lives[9] because it plays an important role in maintaining cognitive function.[10]

I know that a lack of curiosity made me feel even lonelier at a very difficult time in my life, my divorce. My squadron knew that my husband and I were splitting, but they offered the usual pat on the back and surface-level comfort that we all say when something bad happens to someone else: "Let us know if you need anything." I could sense how uncomfortable it was for them. I mean, it was really uncomfortable for me, too, because I wasn't sure how to be vulnerable around them. I was embarrassed and ashamed, and the less that people witnessed my emotions about the situation, the better, or so I thought.

It caused a disconnect in the community that got so bad, I had to force myself to go to work and do my job. For months. I self-isolated because it was very hard for me to share my struggles and ask for help. I was lucky, though, because a few friends stepped up; they insisted that they come over to my apartment, sit in my living room with me, drink wine, and ask very pointed and uncomfortable questions about how I was feeling about my divorce. I didn't seek them out; they came to me. Yet their curiosity and care helped me be vulnerable and soon, we had a little community and I felt less alone.

When curiosity meets vulnerability, people can't help but become connected. A viral *New York Times* Modern Love column showed how.[11] In 2015, a professor shared the story of how she set out to test psychologist Arthur Aron's success in making two strangers fall in love in his laboratory through a series of increasingly intimate questions. She re-created the circumstances, only the subjects this time were a university acquaintance and her, and the location was a bar. They took turns asking Aron's thirty-six questions, starting with, "Would you like to be famous?" through "What is your most terrible memory?" up to "Of all the people in your family, whose death would you find most disturbing? Why?" And wouldn't you know, they fell in love.

In Aron's original research from 1997, he said that the "goal of our procedure was to develop a temporary feeling of closeness, not an actual ongoing relationship."[12] If a series of questions can cultivate intimacy and closeness between two strangers in a lab or two acquaintances in a bar, imagine what they could do when you're deliberately trying to build a community of wingmen.

You don't have to follow the list of thirty-six questions from the study. You can formulate your own questions through curiosity. *Harvard Business Review* recommends harnessing the power of asking good questions by ensuring you remain casual instead of overly formal, sprinkle in open-ended questions, and use follow-up questions to show you're listening and interested.

THE FLIPSIDE

Curiosity and vulnerability are not only important for finding your community, but also for being a good leader. When I failed to correctly drop the laser-guided bomb in Afghanistan, the director of operations quickly recognized my errors in the debrief. But he went beyond that and dug in with curiosity. Soon, he uncovered that there was something else I had been wrongly focused on that led to my mistake.

He could have shouted at me that I hadn't followed orders and stormed out, leaving me red-faced and defeated, but he didn't. Instead, he asked a series of questions that led to new solutions for future missions while making room for me to learn and grow as a combat pilot. He kept a level head and focused on the bigger-picture goal of developing the best pilots. When I became an instructor pilot soon after, I'd keep this in mind when my students made their own mistakes, and I'd avoid making it feel like an interrogation of a suspect. First, I used curiosity, asking open-ended questions to get to the bottom of what they'd been thinking when they flubbed in the cockpit. Then I brought in vulnerability. I made sure they knew that there was never a flight where I, as an instructor, flew perfectly, and I shared my own errors and what I was thinking when I made them. That, in turn, created a safe space for my students to share their mistakes—and this is the important part—without shame.

It was vulnerability, not shaming the person who made the mistake, that allowed us to become as good as possible at our craft. The debrief is a form of accountability, and it's a pillar in the fighter pilot culture, crucial to creating a tight-knit and high-performing community.

Though America loves a lone wolf who gets by on "rugged individualism," humans, like wolves, are pack animals. We thrive in community, even if the comment threads on X (formerly Twitter) might make it seem otherwise. There are several benefits to community, including a sense of security that help is never far, opportunities to learn, empathy, and well-being.

Over the years, public health research has found that anywhere from 40 percent to more than 80 percent of our health and wellness can be attributed to social factors,[13] and a lack of social connection can increase the risk of premature death.[14] In fact, the famous Framingham Heart Study, which tracked the cardiovascular health of several generations of participants over decades, found that happiness can spread along social networks, adding support to ongoing research that has found people with strong social supports tend to be healthier than lonely and isolated people.[15]

Community is also important for mental health, providing purpose, support, and a sense of belonging. This is so fundamental to our being, the Mayo Clinic said that without it, "we would live solitary lives, only coming together for procreation then quickly kicking the children out of our lives as soon as they could walk. We would have no families, communities, or organized government."[16]

I shudder to think what flying with the Thunderbirds would have been like without a sense of belonging. We worked well together because we were a team that trusted one another to fly inches apart and perform death-defying maneuvers. We needed to feel like we belonged in order to maintain that level of trust. Without belonging? At best, it would have been lonely. At worst, dangerous. People need people, and it's our connectedness that helps us overcome loneliness to find a reason to get out of bed every day. I discovered quickly how much connection we can create with a good Instagram page and a LinkedIn following, if you do it right. It's no substitute for being IRL, in real life, but it sure does bring together people with common interests from all over the world really well.

I didn't keep my social media channels going while I was still on the Thunderbirds for the likes and comments. It was for the personal messages I received. One young woman, straight out of ROTC, had just received a pilot slot for training in Columbus, Mississippi, which was where I trained. Like me, she was a non-technical major, meaning she didn't have an aeronautics or engineering degree. Like me, she would begin pilot training with zero flight hours because she didn't even have a private pilot's license, the most basic of flying permissions. She was pretty much me back in the beginning of my career, and she was really worried that she'd gotten in over her head. She reached out to me because she saw videos of me in my feed, flying an F-16 upside down. I shared with her time-management tips for the grueling year of pilot training, and I explained how I dealt with the stress.

Eighteen months later, she messaged me again.

"You probably don't remember me..." Oh, but I did. She'd just had her Drop Night, when soon-to-be newly minted pilots find out which aircraft they'd get. Like me, she got an F-16.

There aren't many women flying in the Air Force and even fewer in fighter aircraft, and yet, social media brought us together, automatically folding her into my community. But social media hasn't just helped me build my military wingmen community, it's also brought me new friends and experiences, like the chance to become a Swiftie.

I met my friend Nicole online. We clicked "like" on each other's content and exchanged messages from time to time. I was missing camaraderie because most of my military friends had moved out of Las Vegas. Nicole was an aviation enthusiast and outdoors activity fan who happened to have an extra ticket for the Taylor Swift concert that she needed to sell. Instant friends.

That's because real people are behind the avatars we see on social media. Sure, there are trolls and bots and plenty of jerks, but there are also real people seeking connection and inspiration. I'm very mindful about the type of content I share so that I can build the kind of online community I'd want to have if we were in one another's living rooms. It's a positive space where I avoid polarizing topics like politics and professional football. When that rare troll rolls in to crap on my feed, I call them out, pointing out that perhaps that post or my feed simply isn't for them. Move along.

Whether you're connecting with someone online or IRL, it's important that you make genuine connections instead of choosing image over substance, the perfect Insta vs. the imperfect relationship.

BerkeleyExecEd provides five tips for cultivating solid connections:

1. **Know what you need from your relationships.** That way you can weed out the ones that don't help you achieve your connection goals.
2. **Recognize different levels of connection.** That said, some people don't want to "go deep," preferring to keep their relationships at surface level. Not everyone needs to be your ride-or-die, and every community is made up of different types of people.
3. **Be patient and proactive.** Building connections takes time and effort.
4. **Be yourself.** Authenticity leads to genuine connection.

5. **Reflect on your interactions.** When the interaction feels good, try to cultivate more like it.[17]

After several years of building my own online communities, I'd add, "Be vulnerable," to that list. Nobody demonstrated that better than Taj, who was unapologetic when he reached out for help to escape from Afghanistan. If he could do it despite cultural differences and years having passed, surely I could add vulnerability to my social media posts. When I do, it often seeps into the comments, like when a sales executive admitted in a comment under my inner critic post on LinkedIn that he'd let fear get in his way. Or when a former pilot shared in my Instagram feed that he was struggling with his new life as an entrepreneur. I may get my fair share of mansplainers advising me how to fly my own jet, but mostly, we're all in there for the inspiration, the education, and the cool flying videos.

From my very first post, my goal for my social media accounts was to connect with people because it feels really good when you give someone a gift that they're really excited about. I knew all about this kind of community from the autograph line after Thunderbird air shows, where I'd be able to share behind-the-scenes information that the audience craved. Many were excited just to get one simple question answered.

It was about more than the jets. I was a real person they could connect with about the very thing that excited and interested them enough to stand in line in the sun after a long day in order to meet a fighter pilot. I wanted to re-create that feeling of excitement and curiosity online, where I could supply what the official feeds for the Air Force and the Thunderbirds couldn't: what it feels like to be a fighter pilot, in the cockpit and on the ground.

Since then, I've gained followers and I've lost followers as I pivoted my career, my goals, and my content. But I've learned that if you do it right, your community finds you. Always. They were there for me when I left active duty to launch a new career, and when I wrote my first book, a children's picture book called *Upside Down Dreams*. They stepped up to help connect me to the people I needed to know to make my new goals and dreams come true, putting a lot of resources and trust behind me. They checked my six. None of that would have happened if I'd never evolved from that lone wolf going it alone.

I have been shown so many times that others are willing to be my wingman for me, reaching into their networks and giving me the most valuable resource of all: their time. I learned through repeated exposure to great people showing up over and over again, and I believe in my heart, that it's way worth it to build your community of wingmen. But that also means being other people's wingman, too.

WE ARE ONE ANOTHER'S WINGMEN

When Taj reached out to his former fellow pilots for help escaping Kabul, I was in the middle of my air show season with a lot going on. I could easily have left the chat with a "Dang, that sucks. Good luck with that," and gone on with my life. So could have my colleagues Ryan and Eric, but instead, they took turns sleeping just four hours a night so that there was always one of them awake, in different time zones, to work leads with government agencies. They had jobs and families, but they went all in to help Taj and his family. Taj's wingmen were willing to make it a priority because

THE FLIPSIDE

the gravity of the situation outweighed anything happening in our own lives, and when you're a wingman, you show up and help.

Taj's situation was dire and extreme, and most of the time, your community doesn't need that kind of urgent support. And I'm not saying you should abandon your own needs or ditch your family to always run around helping other people. That's a surefire way to burn yourself out and cause resentment among those closest to you. But when you build your community of wingmen, it becomes clear who's in it and who has your back and vice versa.

As for my community, I understand that I can't answer every request for mentorship or assistance because I don't have the time or energy. Nobody does. I have to use my resources wisely. When people have something in common with me, like the young fighter pilot heading into training, or we've been through something important together, like Taj, or when we share similar visions, whys, and dispositions, that's when we can consider building relationships that grow into community. And when you have a community, someone will check your six for you. Be sure to check theirs, too.

I'm grateful for my wingmen, even the ones who were just in my life for a short time. When I was in pilot training, an instructor with a fighter jet background fueled my motivation for months to come with just a few words. After most uneventful flights, our instructors would tell us we did a good job, hand us our grade sheets, and move on to their next flight. This time, though, my instructor took a moment to tell me something that felt incredibly exciting to me:

"That was a great flight. I think if you continue to work as hard as you're working, and continue to take instruction as well as you do, you're going to walk away from this program with a fighter aircraft."

It was a short interaction with a very long impression, and it kept me motivated for the remainder of the program. As he predicted, I did get a fighter aircraft.

It's not like he simply said, "Good job! Your dreams are going to come true." He took the time to say a bit more that gave me a few conditions and a bit of direction to set me up for success. He was being a good teacher and therefore, a good wingman to my fledgling career. Never underestimate the power of short but important interactions on the people in your community. Ultimately, that's what it means to check their six.

Scriptside:

- Going it alone.
- Making assumptions and never asking questions. (You don't want to look dumb, right?)
- Looking and acting tough.
- Letting people figure it out for themselves.

Flipside:

- Being open to help.
- Being curious.
- Being vulnerable enough to connect, inspire, and lead.
- Being a good wingman.

CHAPTER TWELVE
You're Saying There's a Chance

A fighter jet doesn't care if you're a man or a woman, but the people on the ground at an air show do, and for five years, there were no female Thunderbirds for little girls to (literally and figuratively) look up to. Five years of girls not seeing themselves represented in that role is a long time, especially considering it had taken more than fifty years for the Thunderbirds to bring on their first female pilot. So when the final call for Thunderbird applications hit my computer's inbox, the gap that needed to be filled to inspire the next generation of aviators was on my mind.

I'd seen the Thunderbirds myself for the first time at an air show in Columbus, Mississippi, when I was waiting to transfer to Luke Air Force Base in Phoenix to learn how to fly F-16s. Back then, I was a hotshot, a recently graduated, soon-to-be fighter pilot assigned to fly the same kind of jets as the Thunderbirds, and yet I was too shy to introduce myself to the only female Thunderbird pilot at the time, Number Eight, the Narrator with the call sign "Mother," at a reception at the officer's club.

"Go tell her you're gonna fly Vipers," my friend Kristina, who was in the class behind me, said. But I was too intimidated, choosing instead to eye Mother (real name, Kristin Hubbard) up from across the club all night. Years later, I'd finally meet her, and she'd become a valuable mentor to me, but that would be after I'd logged several hundred hours in the F-16 and stopped viewing other female fighter pilots through a lens of competition or intimidation.

By the time I was mulling over applying to the Thunderbirds, I had been in the Air Force for nine years. I was wrapping up my assignment in Fort Worth, Texas, and I was getting ready to serve as an instructor pilot at Holloman Air Force Base in New Mexico, teaching new F-16 pilots how to fly the jet. Though I'd been asked before if I wanted to apply for the Thunderbirds, I'd long felt like I lacked the experience and the courage. Yet there were three openings: Thunderbird Number Three, which is the Right Wingman; Number Six, which is the Opposing Solo; and Number Eight, the Narrator who coordinates and announces at air shows. This time, I met the requirements, such as minimum flight hours and career timing, and I knew my age, active duty service, and solid track record would be viewed favorably. I was in the sweet spot to qualify, only I'd missed two prior application emails and now the deadline was fast approaching.

I was about to make a bold move. A very bold move.

I approached my boss and told him that I wanted to apply for the Thunderbirds and oh, by the way, the deadline is next week. He could have said, "No, wait until next year." He could have reminded me that Holloman was expecting me, and he could have pointed out that I'd never expressed interest in becoming a Thunderbird in our career mapping meetings over the years. Instead, he asked, "So you want to apply for the Thunderbirds?"

"...I think so." It wasn't a movie-script-worthy "Sir, yes sir!" or "It's been a dream of mine since childhood." Just "...I think so."

"You would be great for that job."

For the next five days, he helped me jump through all sorts of hoops to get the forty-page application done. I needed letters of recommendations from generals and photos of me in my service dress. I needed records pulled and pages and pages filled out. I knew I should have brought it up months earlier, but here I was, scrambling to apply for one of three open positions, in less than a week.

I had become the bold person who does the hard thing. It started back at my squadron in Fort Worth, when I'd be asked on Mondays, "What crazy thing did you do this weekend?" Rock climbing. Hiking. Bungee jumping. My squadron of mostly older pilots with families were living vicariously through me.

So when the Thunderbirds application came across my desk, it felt so much less scary to take the calculated risk of applying than it would have years earlier. Besides, there were little girls—and boys—who needed to see someone like me in that role.

I got the application in just in time and, after a series of interviews, I was hired by the Thunderbirds, first as Opposing Solo and later as the Lead Solo. I'd taken everything I'd learned, in the cockpit and out, and I gave myself the chance of a lifetime. In the process, maybe it would inspire other girls to someday become a Thunderbird, too. I'm saying there's a chance.

NOT DUMB AT ALL

If you've ever seen *Dumb and Dumber* you may remember when Jim Carrey's character asks his unrequited love interest, played by Lauren Holly, what the odds are of them ending up together.

"Not good," she replies.

"You mean, not good like, one out of a hundred?"

"I'd say, more like, one out of a million."

He takes a deep breath and then slowly begins to smile, broken tooth and all. "So you're telling me there's a chance."

The odds of me flying for the Thunderbirds had been a long shot. Of all the kids who dream of flying fighter jets, a few go on to get accepted into Air Force flight school. Of those, even fewer finish, and of those who do, only one or two in each class get assigned a fighter aircraft. Of those, a few want to become a Thunderbird and are willing to take on the added burden of the intense flying schedule and the PR responsibilities that come with it. A few of those have also clocked enough flight hours to qualify and their career timing is right, and even fewer actually apply. (Some of them wait till the last minute. Ahem.) There are just eight spots for fighter pilots on the Thunderbirds, and of those, just six of them get to perform in air shows. Of those, just the Solos get to do the head-on opposing passes with 1,000 miles per hour of closure. And of those, only the Lead Solo performs the Calypso and Reflection Pass inverted, when the jets appear to be flying belly-to-belly or tail-to-tail, with one of them upside down. Out of all the demonstration pilots to fly for the Thunderbirds since the program's inception more than seventy years ago, just six have been women.

If you'd told me when I was watching those fighter jets take off back when I was in ROTC that I'd be the one to get to do all that,

THE FLIPSIDE

I might not have believed you. Then again, if you had said there's a chance, I probably would have taken it, but first, I would have to overcome my own fears and look on the flipside.

"You're saying there's a chance" is about being willing to do the thing that might move you toward your audacious goal or big dream, even if it feels uncomfortable, scary, or uncertain. You're going to feel resistance because it's not something you're used to doing. It may even make you feel like an imposter. But looking at things from another vantage point can help you get through just about anything. And I would need it, because my role on the Thunderbirds wasn't just mentally challenging. It was also hell on my body.

FROM DENIAL TO PERFORMANCE

During my first year on the Thunderbirds, I missed three shows because I couldn't get off the floor without my back spasming. All those G-forces take their toll, and I started suffering from chronic back pain just a month into my first air show season. I was out flat.

It was very noticeable when a jet was missing from a performance, when five jets show up instead of six. Everyone expected us to show up at 100 percent capacity, all the time, and that's hard to do even when you feel well. It's already a difficult assignment. That's why it's usually two years long for the officers, forcing a 50 percent team turnover every winter. The energy drain that comes with being on the road 240 days a year, combined with the responsibilities of being a public figure can burn you out quickly. Now I had the added burden of cervical and lumbar pain running underneath it all. By the time I left the Air Force, my body was so beat up, a VA doctor asked me if I'd been in a car accident because the damage to my vertebrae was something she'd normally seen in someone over sixty years old or in a patient who'd suffered some sort of trauma.

Flying fighter jets is just plain hard on the body. Normally, your head weighs about eleven pounds, but in a cramped cockpit at nine Gs, it can feel nine times heavier. When was the last time you lifted one hundred pounds with your neck? That hurts. Even with all our training, plenty of fighter pilots end up with neck and back pain, because when you're young and dumb and you're dogfighting, your drive to win surpasses the logical part of your brain that says, *Don't move your head under Gs.* But if you lose sight, you lose the fight, as we say. So we end up twisting our necks to look for bad guys under the pressure of up to nine Gs. Even in my early twenties, my neck used to make a sound like popping popcorn after I landed from our dogfight training.

By the time I flew for the Thunderbirds, all those Gs had taken their toll. One day, I drove home from a practice flight at Nellis and pulled my Volkswagen into my garage and parked. But my legs wouldn't get out of the car. I couldn't lift them or else my back would spasm. I was stuck there, holding on to the steering wheel, having managed to get one leg out of the car and one still in, wondering if I was going to have to honk the horn for help. At night, my husband would have to help me lift my legs into bed.

To endure all this, I was given steroid shots, eight hundred milligrams of ibuprofen, and lidocaine patches. I had been in the best shape of my life. I'd run two marathons and a 50K trail ultra race. I'd climbed multiple mountains and I pushed myself in the gym. But now, I was putting my body through the ringer, on land and up in the air, and my body was finally like, "Nope. Not doing it."

THE FLIPSIDE

During an air show, the Solo flight profile is the highest G of any position on the team. When the Solo flies the Max Turn Half Cuban, we're subjected to seven to nine Gs for a full 360-degree turn just 150 feet above the ground. Remarkably similar to the move that earned me my call sign, this maneuver impresses the audience, but it's hard on the pilot's body. For three years, that pilot was me, with two air shows and multiple practices per week during the show season, and practice shows up to ten times per week during the winter training season.

A few previous Lead Solos admitted to me that they did all sorts of crazy things to get their bodies to work in and out of the jet, but for the most part, nobody talked about it. If I'd brought it up beyond a small group that I trusted and sought out imaging and medical treatments, I'd put myself at risk of going DNIF—Duties Not Including Flying. I'd be out of the jet. For most pilots, this is our greatest fear. We'd go through all that training and hard work, becoming part of the community, only to have some medical issue sideline us. Usually, it's a temporary DNIF for issues like a sinus infection or a broken wrist. But this? I'm not sure I would have been able to stage a comeback.

My second year in the Thunderbirds was 2020, and the pandemic cut our usual forty-show schedule down to six locations. That was easier on my back. Plus, I'd learned how to adjust my lifestyle (no more deadlifting for me), and soon I felt a bit better. As the end of my two-year assignment approached, I began to make plans outside the F-16 and outside the Air Force. Only, it wouldn't be the end after all. Not yet.

I started to hear rumblings that we were going to be asked to stay a third year because 2020's reduced schedule meant fewer flight hours for everyone. And that meant that my teammates who were in their first year wouldn't have the depth of experience that comes with a normal show season, but they'd be responsible for instructing the new hires as those of us on our second year departed the team. The Thunderbird mission already comes with a lot of risk. Handing off a new team to be run by pilots lacking the normal depth of experience brought on more risk than the Air Force was willing to take.

I handled the impending news by pretending it wasn't real. I was a newlywed with a young stepson, and I'd already been away two-thirds of the year early in our marriage. Plus, my back and neck still hurt. But I couldn't be the only one to leave the team. Everyone else had planned to leave the Thunderbirds for another active duty assignment, while I was planning to leave the military altogether six months later to become a civilian. Now that dream would be another eighteen months away. But I wouldn't let the team down. I wouldn't screw over my friends by increasing their odds of bad outcomes, and so, I agreed to stay.

I struggled to reconcile my feelings with what the public thought. Once our third year was announced, my social media accounts filled with congratulatory comments. In interviews, I couldn't share my real feelings as I hadn't yet reached a point of acceptance. I would get so mad at the disconnect between the public-facing image and what I was actually experiencing, because I felt like I was once again playing dress-up, putting on a smile and nodding my head along with the reporter who was congratulating me on my extended tour. Then I would feel guilty for not wanting what so many people saw as a dream job. How could I be so ungrateful?

Turns out, my third year on the Thunderbirds was the best year I had. Once I got myself from denial to acceptance, I began to

THE FLIPSIDE

thrive. I'd learned how to manage my pain by adjusting my workouts and mobility training. On Sundays, I'd turn down all sorts of cool invitations—things like five-star dinners and backstage concert tickets that were frequently offered up by our shows' hosts—so that I could rest my body and prepare to fly the next day. I'd pack myself healthy, premade meals and send them in a cooler on the C-17 that transported the Thunderbirds' gear and personnel so I wouldn't grab just anything off a food cart at air shows. I got better at prioritizing my battery, and I found my footing in my public persona so that it didn't drain me as much along the fence and on camera.

In other words, I inverted my perspective and discovered the flipside of a tough situation was actually its best side. You can invert your perspective and turn fear into your superpower for whatever it is you wish to do.

As Thunderbird Five, I spent a lot of time flying upside down. So the number five on my aircraft, as for all Lead Solos, was upside down. That way, it appeared right side up for much of the show. Even the number five embroidered on my flight suit was upside down as a nod to the Lead Solo's many maneuvers spent inverted. But that's not all. All number fives that appear in our pamphlets and line-up cards, which listed our start times and maneuvers, were also upside down. I even learned to write my fives inverted.

So when I say I am committed to looking at things from the flipside, I really mean it. When you turn things over, it allows you to see them from another perspective. When you do it with adrenaline racing through your system and that uneasy feeling in your stomach, that's when you know you're using your courage to be bold. And when you're willing to be bold, you inspire others to do the same. But the flying was only part of it. More than anything, I had to learn to be bold on the ground.

MICHELLE CURRAN

IT TAKES COURAGE TO BE BOLD

Flying for the Thunderbirds was hard at times. We had to learn new maneuvers and fly exceptionally close to one another, and there wasn't much room to have a bad day. But, for me, the most consistently hard thing to do was to show up and be "on" for the fans. For many of them, this would be the only time they'd meet a Thunderbird pilot. It's kind of like being a character at Disney World. Cinderella might have period cramps, but she's got to put on that smile and talk to the little girl who's come all that way to see her.

For a while, I didn't thrive in that setting. I wasn't the type to seek out the limelight, and yet I showed up, and I'm so glad I did because it meant that my boldness led to inspiration for other people. As I've said, to inspire is a core value of the Thunderbirds and it's among my top three, and for good reason. *The Collins Dictionary* explains, "If someone or something inspires you to do something new or unusual, they make you want to do it."[1] Imagine having the kind of influence that your actions make someone else want to do something new and unusual. You can, by being bold.

Not all inspiring actions are high profile, but the Thunderbirds surely provided those opportunities. That was how I wound up on the couch on *The Kelly Clarkson Show* next to a seven-year-old girl dressed in a flight suit just like mine. Amelia was obsessed with becoming a Thunderbird ever since her dad, a pilot, took her up flying when she was just a toddler. In fact, both her parents are pilots.

Amelia and I had met very briefly at a Thunderbird air show in

2019, but Kelly brought us together on her show in LA for a special "Whole Lotta Woman" series about women doing inspiring things. Amelia had no idea I was on set when she told Kelly she was inspired by me, explaining that there's a Thunderbird "who's upside-down flying, and she's the Lead Solo, and her name is Michelle." When Kelly introduced the "surprise," Major Michelle Curran of the Air Force Thunderbirds, Amelia jumped the arm of the couch and ran into my arms. I picked her up, and then her feet, which were in her trademark cowboy boots, swung back and forth, as I carried her over to the couch so we could sit together. I told her that my job was to inspire people, but that she'd inspired me with her determination and dedication to her dream.

A week later, the show brought her out to Nellis Air Force Base, where she got to spend a day with me at work, even posing next to an F-16 with her name on the side. Someday, she'll probably have her own F-16 for real. Her family has exposed her to aviation and she's been flying aerobatics since she was six, performing aileron rolls with her dad, one for each year of her age, until they reached the maximum allowed.

The show also provided inspiration for other kids who, like me, weren't from a flying family, didn't live near a military base, and didn't know any fighter pilots. Maybe it will inspire kids, especially girls, to pursue a career in the cockpit.

Because it showed them there's a chance.

Throughout this book, I've shared the beliefs, concepts, and lessons that helped me overcome imposter syndrome to become a Thunderbird pilot. Though I'm no longer in the cockpit, I still use many of these to achieve my goals and live a good life. Here's what to remember as you plan your own flight path:

GO TO WAR WITH YOUR INNER CRITIC

Self-doubt and fear originate with your inner critic, which takes delight in judging and criticizing you. I've identified Five Inner Critics, each with its own goals: the Fear Critic, the Comfort Critic, the Friend Critic, the Reputation Critic, and the Wrong Critic. You must go to war with any and all of them, first by knowing your loadout, the things that make you feel most like yourself and most confident. Then, by identifying your draws, which are your next-level hopes and dreams. Say them out loud and write them down. Next, name your wingman or wingmen. When you've created your battle plan, fight's on… start taking action! Your inner critic is just a voice, not a verdict.

EVERYONE HAS A CALL SIGN

What's your version of nearly G-LOCing? What lesson have you needed to learn, perhaps more than once, and have you held yourself accountable for your mistakes? I'm not talking about shaming yourself or letting someone else shame you into behaving differently. I mean debriefing the situation and figuring out how to handle it better next time, issuing a course correction. Then look for the "hell yeahs" and head straight for them. Mistakes can define you or refine you.

STOP PLAYING DRESS-UP

After your inner critic starts bullying you, it's easy to get a case of imposter syndrome. You feel like a fraud and get stuck, so you play it small. But you're not a fraud. You're a beginner, that's all. You're new. Even if you've been at your job or in a role for a long time, it constantly evolves, requiring new skills that you must work on to move forward. You'll need some authenticity and vulnerability to acknowledge your beginner's mindset and then seek out the information you need to get unstuck. Stop dressing up and start showing up.

REMEMBER TO WIGGLE YOUR TOES

If you find yourself frozen in fear over a situation, wiggle your toes to take your attention off the problem and into your body. If you don't reset yourself like that, you could wind up stuck with anxiety-inducing brain waves that don't serve you or your mission. Use the pause created by that physical cue to choose how you want to respond. Then, don't go it alone. Assemble your Dream Team, using the wisdom of others to help you keep calm and carry on. Finally, don't fight the jet, because trying too hard to control something that doesn't need controlling is a waste of your time and energy. Find your calm in chaos.

TAKE ONE MINUTE, ONE HOUR, ONE MONTH FOR CHANGE

Whatever makes up your version of a bird strike, something unexpected that can set you off course or worse, be sure to maintain control, analyze the situation, take proper action, and "land" as soon as conditions permit. If you can train yourself to respond instead of react, you can take control of situations that usually take control of you. Take one minute to pause, one hour to reflect, and if there's time, one month to transform to better prepare for the next incident. Finally, use a growth mindset to focus on what you learned to respond better next time. Hit pause to turn panic into purpose.

DON'T GET STUCK IN THE COMFORTABLE MISERY

It's easy to choose misery over change, a sort of "devil you know" decision. But it stunts your growth when really, posttraumatic growth can help you not only overcome the misery but learn from it. Posttraumatic growth helps you become psychologically resilient so that you bounce back faster and better. Some of it just happens, if you let it, and some of it requires a concerted effort on your part. Change is hard work, but without it, it's nearly impossible to make sometimes small, sometimes huge improvements toward your goal. Choosing comfortable misery over uncomfortable uncertainty is what keeps you stuck.

SEEK SILVER LININGS

This is more than puppies and rainbows or vision boards and manifestation. This involves building resilience by taking the 30,000-foot view. When you create a new perspective by acknowledging one good thing that came from the bad (or not-so-great), you train yourself to find gratitude even when everything pretty much sucks. Learn to leverage regret—especially regret for things you *didn't* do—and you'll begin to see the silver linings all around you, if you'd just give yourself the grace to learn from it. Regret finds the cracks; resilience fills them in.

THE FLIPSIDE

FOCUS ON THE NEXT CLOSEST ALLIGATOR

If you try to think about problems that might develop too far into the future, you'll lose focus on the task at hand and risk sensory overload. Use the Yerkes-Dodson law to your advantage to identify the sweet spot between "arousal" and anxiety, where performance peaks. When you get into flow, you maximize your self-efficacy to help you face your fears. Meanwhile, reduce decision fatigue through good habit making, and rethink anything negative through cognitive reappraisal, so that you stop the all-or-nothing thinking, catastrophizing, and "should-ing" all over yourself. Giving in to panic today creates pain tomorrow.

YOU'RE BETTER LUCKY THAN GOOD

Luck doesn't just happen. You can help create it, especially smart luck, which involves strategies to improve your odds of a favorable outcome. Among the ways to achieve that are positioning yourself better, noticing more, and adopting a portfolio view for risk-taking. Learn to connect the dots to make serendipity work for you so that you wind up creating pinch-me moments that make you feel lucky to be alive. And always be ready to pivot. Luck doesn't just happen; it's fought for by those who prepare, pivot, and see the possibilities.

AVOID GET HOME–ITIS

When you're in a hurry to get to the end of the mission, you can start to make mistakes. This is when you're the most vulnerable, and it's best to set up some rules before you get to that point. One rule is to establish your Go/No-gos to mitigate the sunk cost fallacy that tends to creep in far into projects, plans, and goals. Another is to define your why, filtering it through your core values, allowing you to make reasoned decisions that are still true to your heart. If you need some help, create an alter ego to help you grow into who you want to be, Sasha Fierce style. It's a great way to bolster your self-confidence and keep the inner critic quiet while you go for bold things. Decisions made in calm guide us in chaos.

CHECK YOUR SIX

Fighter pilots never fly a mission alone and you shouldn't, either. Create a community of wingmen and you'll never fly alone again. You'll need to be vulnerable enough to accept help and curious enough to dive deeper into your relationships of all kinds. When you build your community, remember that you, too, are a wingman for other people. Step up to be there for them, and chances are they'll return the favor. Solo we build skill, but together we build strength.

YOU'RE SAYING THERE'S A CHANCE

This tongue-in-cheek saying has been so much a part of fighter squadron culture, I almost forgot it came from a movie. It's more than living life with a glass-half-full viewpoint, because that only gets you so far. Too often, we look ahead to our dreams and goals and see only the obstacles stacked against us. We focus on our chances of failure and what failure might look like, letting all those negative feelings take charge. But if you train yourself to look at the flipside over and over, it becomes your norm and soon, you automatically see the positive amid all the negative. It becomes your default. That's when "You're saying there's a chance" comes into play, allowing you to treat the odds that are in your favor and the slim chances with the same boldness and courage and go for it anyway. Boldness is the bridge between slim chances and big opportunities.

FIGHTING FOR INCHES ON THE FLIPSIDE

On the Thunderbirds, I could stare only at the wing of the jet next to me even while our formation was barreling straight toward the ground at five hundred miles per hour. That was the job and flying that close was all about "fighting for inches." See, from the ground, flying formation looks smooth, but up in the air, we pilots are working hard to constantly make small adjustments and tiny bits of progress, even if our arm goes numb, our neck cramps, the sun is in our eyes, we're hot and tired, and we want to give up. This fighting for inches makes for a good air show. The willingness to fight for inches, find my wingman, and create trust are lessons that continued to serve me in the air, but even more so on the ground.

I didn't go from being a college student with big dreams to Thunderbird Lead Solo overnight. Along the way, over the years, I fought for inches, tiny bits of progress, learning from mistakes, adjusting, and making choices both minor and bold. I still do.

Near the end of my Thunderbird career, I told a local TV station in Kansas, "You're going to have big dreams and there's going to be people that doubt you, but that's their issue, not yours."[2] Then I explained that even seasoned Thunderbird pilots have self-doubt and fear and that we get stressed out and overwhelmed. But when you're bold, you can overcome that and you'll build a community to support you. I still believe that.

Not long into my new speaking career, I was invited to Oshkosh, Wisconsin, to the annual conference for the Experimental Aircraft Association, a community of aviation enthusiasts who build, fix, and fly airplanes. I spoke to about two hundred people from the small stage, and then took my seat in the sold-out lunch to see the keynote speaker, astronaut and shuttle commander, Eileen Collins. I looked around at the nearly one thousand people filling the chairs, and whispered to my friend, "I want to keynote this event next year."

The event's theme? Be bold. And I was. That was a big ask so early in my speaking career, especially in a room full of so many accomplished people in aviation. But there was a chance, and I took it.

A year later I stood on that very stage where Eileen had spoken so I could keynote a sold-out lunch and sign copies of my children's book, *Upside Down Dreams*, which had just been published. How's that for a pinch-me moment?

On the final day of Oshkosh, an eleven-year-old girl approached me. She seemed shy and nervous to talk to me and perhaps even sad. She had a question, which she eked out: "What do I do if I say I want to be a pilot but the boys at summer camp say girls can't be pilots?"

Wait, what? Kids still say this stuff? Yes, they do.

"Well, you came to the right place," I assured her. "Let me show you something." I pulled out my cell phone and opened up Instagram. I scrolled through my posts until I found one from my Thunderbird cockpit, where my jet flipped upside down and my braid, which extended from under the back of my helmet, flipped with it.

Then I gave her a coin called an RMO (that's military for "round metal object"), which we give as gifts or as a token for a job well done.

"Sometimes, people don't know what they're talking about if they haven't seen it for themselves," I said. "Next time they say girls can't be pilots, stay focused on your dream and remember that video you just saw of the flying braid. The only thing that matters is if you believe you can do it."

"Okay! Thank you!" She lit up, holding her new coin as she skipped away.

I remember that feeling, that I could be anything and do anything if I just believed it. When I was a little girl, I used to dream that my superpower was flying. Not in an airplane, but in my house, hovering above the ground, up the stairs, and overlooking everyone and everything. But in the end, flying wasn't my superpower. Flipping over fear was.

By the time I met that girl in Oshkosh, I had grown into the fighter pilot—and the person—I was meant to be despite it all and because of it all. I'd looked at the flipside, and, it turns out, that's where I'd belonged all along.

ACKNOWLEDGMENTS

Just like a fighter pilot never completes a mission alone, neither does an author. You wouldn't be reading this book if it weren't for so many people who believed in me, sometimes more than I did.

First, my husband, John, and my stepson, Milo. You gave me the time and space to dedicate so many hours to writing and editing. You listened as I brainstormed and worked through my thoughts and feelings around what to share and how to do it in a way that felt authentic. I love you both more than you even know.

Thank you to my entire family, especially my parents, who have endlessly believed in every wild dream I set my sights on. It hasn't gone unnoticed, and I hope you know that it's the foundation you created for me that has allowed me to do all that I have.

Mango, you have been my confidante through the highs and lows. You've asked hard questions when I needed them, and you provided unwavering support. I'm forever grateful to have you as a friend, a sounding board, and a travel buddy.

There are so many fellow fighter pilots who played a role in the stories you see here. Whether I was struggling or thriving, there were always others teaching, supporting, and challenging me. Many of you are briefly mentioned in these stories and many are not, but you all had an impact: Smokin', Siren, Taboo, Klepto, Metro,

ACKNOWLEDGMENTS

Jik-Do, Mother, Bear, Stump, Flack, Skate, Gumbo, "Jen," "Taj"… and so many more.

Thanks to all my fellow Thunderbirds. We spent more time together than we did with our families, and we trusted each other with our lives. Only those who were there will fully understand the challenges we navigated. From those flying a few feet from me, to my crew chiefs, to the entire support team, you made that assignment what it was in the best way.

Thank you to my countless military friends strewn across the country and the world. The one-off conversations we've had, the unexpected reunions, and the ability to pick back up where we left off after years apart make you the stand-in family I've needed for certain situations. I'm grateful for all of you.

Jen Singer: This book would not have happened without you. The countless hours we spent discussing ideas, iterating, writing, editing, and laughing. I'm grateful to you and I'm proud of what we created together—a book I believe in and a deep friendship.

My agent, Matt Latimer, and the entire team at Javelin, and my editor, Jacqui Young, and the team at Grand Central: You believed in me as I forged a path in a new industry. Thank you for your support, for answering my endless questions, and for giving me the roadmap forward.

Thank you to the team at Heroic Public Speaking for helping me get my message ready for the stage, and everyone who helped me make such a huge career pivot: Destin, for planting the seed that this was possible, and Darren, Melissa, Tom, and David for keeping it all running. I am grateful that St. Mark's School of Texas gave me my first stage, and thanks to FiFi (Nicole) Malachowski, the first woman to fly for the Thunderbirds, for showing me what is possible in the air and on the stage.

ACKNOWLEDGMENTS

I offer my appreciation for everyone we quoted and referenced throughout this book, especially those who worked directly with us. Stefanie Faye, thank you for your expertise and advice as we explored brain science. Katie Boer, thank you for sharing your story and showing me how much a book can change a life.

Finally, I offer my love and appreciation to all of you who have supported me both online and in person. Whether you found me through my early Thunderbird cockpit videos or joined later from a LinkedIn post or this book, I am constantly amazed at how you all show up again and again. Your support helps keep me motivated to continue to share. **THANK YOU!**

ENDNOTES

CHAPTER ONE: GO TO WAR WITH YOUR INNER CRITIC

1 Washington, Lindsey. "How to Make Friends with Your Inner Critic." *Washington Post*, September 20, 2022.
2 "Demographics of the U.S. Military." Council on Foreign Relations, updated July 13, 2020. https://www.cfr.org/backgrounder/demographics-us-military.
3 Verywell Mind. "What Is the Negativity Bias?" https://www.verywellmind.com/negative-bias-4589618.
4 Csikszentmihalyi, Mihaly. *Flow: The Psychology of Optimal Experience*. New York: Harper Perennial Modern Classics, 2008, 2.
5 Blevins-Holman, Col. Grace. "The Air Force Family and How to Find a Wingman." Joint Base Charleston, October 13, 2006. https://www.jbcharleston.jb.mil/News/Commentaries/Display/Article/238653/the-air-force-family-and-how-to-find-a-wingman/.

CHAPTER TWO: EVERYONE HAS A CALL SIGN

1 Columbia University. Columbia Center for Teaching and Learning. "Feedback for Learning." https://ctl.columbia.edu

/resources-and-technology/resources/feedback-for-learning/. Accessed January 26, 2024.

2 Metcalfe, Janet. "Learning from Errors." *Annual Review of Psychology* 68, no. 1 (2017): 465–89. https://doi.org/10.1146/annurev-psych-010416-044022.

3 Moser, Jason S. "How Your Brain Reacts to Mistakes Depends on Your Mindset." Association for Psychological Science, September 29, 2011. https://www.psychologicalscience.org/news/releases/how-the-brain-reacts-to-mistakes.html.

4 Moser. "How Your Brain Reacts."

5 Duckworth, Angela, and James J. Gross. "Self-Control and Grit: Related but Separable Determinants of Success." *Current Directions in Psychological Science* 23, no. 5 (2014): 319–25. https://doi.org/10.1177/0963721414541462.

6 Hanford, Emily. "Angela Duckworth and the Research on 'Grit.'" American Public Media. https://americanradioworks.publicradio.org/features/tomorrows-college/grit/angela-duckworth-grit.html.

7 Duckworth, Angela L., Abigail Quirk, Robert Gallop, Rick H. Hoyle, Dennis R. Kelly, and Michael D. Matthews. "Cognitive and Noncognitive Predictors of Success." *Proceedings of the National Academy of Sciences* 116, no. 47 (2019): 23499–504.

8 David, Susan. "When to Grit and When to Quit." LinkedIn, April 3, 2017. https://www.linkedin.com/pulse/when-grit-quit-susan-a-david-ph-d-/.

9 "Why the Brain Makes Mistakes." *Neuroscience News*, December 8, 2015. https://neurosciencenews.com/mistakes-neural-networks-3235/.

ENDNOTES

10 Shen, Libing. "The Evolution of Shame and Guilt." *PLOS One* 13, no. 7 (2018): e0199448. https://doi.org/10.1371/journal.pone.0199448.
11 TED. "Listening to Shame: Brené Brown: TED." YouTube, video, March 16, 2012. https://www.youtube.com/watch?v=psN1DORYYV0.
12 TED. "Listening to Shame."

CHAPTER THREE: STOP PLAYING DRESS-UP

1 Clance, Pauline Rose, and Suzanne Ament Imes. "The Imposter Phenomenon in High Achieving Women: Dynamics and Therapeutic Intervention." *Psychotherapy: Theory, Research & Practice* 15, no. 3 (1978): 241–47. https://doi.org/10.1037/h0086006.
2 Clance and Imes. "The Imposter Phenomenon."
3 Li, David K. "'Loser' Gaga Bares Her Pain." *New York Post*, April 26, 2011. https://nypost.com/2011/04/26/loser-gaga-bares-her-pain/.
4 Corcoran, Barbara (@barbaracorcoran). "Who doesn't suffer from imposter syndrome?..." Instagram, February 25, 2020. https://www.instagram.com/p/B9AzPs3nhmA/?igshid=1kxw78wky97ra.
5 Dishman, Lydia. "How to Stop Feeling like a Fraud." Fast Company, March 26, 2014. https://www.fastcompany.com/3028084/how-to-stop-feeling-like-a-fraud.
6 Grant, Adam (@AdamMGrant). "Imposter syndrome isn't a disease…" X (formerly Twitter), September 9, 2022. https://x.com/AdamMGrant/status/1568292869936488448.

ENDNOTES

7. Tulshyan, Ruchika, and Jodi-Ann Burey. "Stop Telling Women They Have Imposter Syndrome." *Harvard Business Review*, February 11, 2021. https://hbr.org/2021/02/stop-telling-women-they-have-imposter-syndrome.
8. Roca, Xavi. "The Three P's of Optimism." Xavi Roca (website), September 15, 2020. https://xaviroca.com/en/the-three-ps-of-optimism/.
9. Henry, Todd. "Overcome Imposter Syndrome." *The Accidental Creative*, podcast, 2023.
10. Zillgitt, Jeff. "Day 61 Without Sports: Revisiting Michael Jordan's Baseball Career with Fresh Eyes." *USA Today*, May 15, 2020. https://www.usatoday.com/story/sports/nba/2020/05/11/michael-jordan-baseball-career-the-last-dance/3105947001/.
11. Hawley, Larry. "Michael Jordan Signed with the White Sox 25 Years Ago." WGN Morning News, February 7, 2019.
12. Castrovince, Anthony. "The Real Story of M.J.'s Baseball Career." MLB.com, February 7, 2024. https://www.mlb.com/news/featured/michael-jordan-the-real-story-of-his-baseball-career.
13. Robbins, Kendall. "These 4 Words Will Silence Your Self-Doubt: A Life-Changing Conversation." *The Mel Robbins Podcast*, August 17, 2023. https://www.melrobbins.com/podcasts/episode-93.
14. Rock, David, and Heidi Grant. "Why Diverse Teams Are Smarter." *Harvard Business Review*, November 4, 2016. https://hbr.org/2016/11/why-diverse-teams-are-smarter.
15. American Psychological Association. "Authenticity." *APA Dictionary of Psychology*. https://dictionary.apa.org/authenticity.
16. Brenner, Grant Hilary. "Sacrificing Authenticity on the Altar

ENDNOTES

of Narcissism." *Psychology Today*, August 20, 2018. https://www.psychologytoday.com/us/blog/experimentations/201808/sacrificing-authenticity-the-altar-narcissism.

17 "Authenticity." *Psychology Today*. https://www.psychologytoday.com/us/basics/authenticity.

18 Leone, Dario. "You Gotta Be Shitting Me! The Story of the First U.S. SAM-Hunters in Vietnam." The Aviationist, March 13, 2014. https://theaviationist.com/2014/03/13/wild-weasel-f-100/.

19 "Daring to Be Vulnerable with Brené Brown." University of Minnesota. Taking Charge of Your Wellbeing. https://www.takingcharge.csh.umn.edu/daring-be-vulnerable-brene-brown.

20 Maciejewski, Justin, and Rob Theunissen. "How the British Army's Operations Went Agile." McKinsey & Company, October 31, 2019. https://www.mckinsey.com/capabilities/people-and-organizational-performance/our-insights/how-the-british-armys-operations-went-agile.

21 Sinek, Simon. "Let's Talk About Vulnerability." YouTube, video, October 12, 2022. https://www.youtube.com/watch?v=kvakdHlrFTM.

22 Sinek. "Let's Talk About Vulnerability."

CHAPTER FOUR: REMEMBER TO WIGGLE YOUR TOES

1 Jha, Amishi P., Elizabeth A. Stanley, Anastasia Kiyonaga, Ling Wong, and Lois Gelfand. "Examining the Protective Effects of Mindfulness Training on Working Memory Capacity and Affective Experience." *Emotion* 10, no. 1 (2010): 54–64. https://doi.org/10.1037/a0018438.

2 Bengtsson, Christina. "Mindfulness vs. Focus: What's the

Difference?" LinkedIn, November 12, 2019. https://www.linkedin.com/pulse/mindfulness-vs-focus-whats-difference-christina-bengtsson/.

3 TEDx Talks. "The Art of Focus—A Crucial Ability: Christina Bengtsson: TEDxGöteborg." YouTube, video, February 9, 2017. https://www.youtube.com/watch?v=xF80HzqvAoA.

4 "Brainwaves Explained—Alpha, Beta, Delta & Theta—What These Brainwaves Mean for Your Brain Health." Clarke Bioscience, February 27, 2023. https://clarkebioscience.com/brainwaves-explained-alpha-beta-delta-theta-what-these-brainwaves-mean-for-your-brain-health/.

5 "Brainwaves Explained."

6 "What to Know About Gamma Brain Waves." WebMD, August 16, 2023. https://www.webmd.com/brain/what-to-know-about-gamma-brain-waves.

7 Sinek, Simon. "Nervous vs. Excited." YouTube, video, May 16, 2018. https://www.youtube.com/watch?v=0SUTInEaQ3Q.

8 Capture Your Flag. "Simon Sinek on Training Your Mind to Perform Under Pressure." YouTube, video, January 9, 2014. https://www.youtube.com/watch?v=GBF9xXhSFRc.

9 Wilkinson, Stephan. "The Day an F-106 Supersonic Fighter Landed Itself in the Middle of a Field." HistoryNet, February 3, 2021. https://www.historynet.com/f-106-the-cornfield-bomber/.

10 Copp, Tara. "What We Know About the Marine Corps F-35 Crash, Backyard Ejection and What Went Wrong." Associated Press, September 19, 2023. https://apnews.com/article/f35-crash-military-marines-plane-0a99d551aeff9eeab1a105215d7f203d.

11 Bazail-Eimil, Eric. "Wreck of Missing F-35 Found as Marines Ground Flights." *Politico*, September 18, 2023. https://www

.politico.com/news/2023/09/18/missing-f-35-marines-pilot-flights-00116643.

12 Imperial War Museums. "Keep Calm and Carry On: The Truth Behind the Poster." YouTube, video, November 3, 2017. https://www.youtube.com/watch?v=s16YLhKzfNI&t=223s.

13 "Keep Calm and Carry On: The Truth Behind the Poster."

14 Hatherley, Owen. "Keep Calm and Carry On—the Sinister Message Behind the Slogan That Seduced the Nation." *The Guardian*, January 8, 2016. https://www.theguardian.com/books/2016/jan/08/keep-calm-and-carry-on-posters-austerity-ubiquity-sinister-implications.

15 Jack, Malcolm. "How We Made the Keep Calm and Carry On Poster." *The Guardian*, April 20, 2020. https://www.theguardian.com/artanddesign/2020/apr/20/how-we-made-keep-calm-and-carry-on-poster.

CHAPTER FIVE: TAKE ONE MINUTE, ONE HOUR, ONE MONTH FOR CHANGE

1 Taylor, Jim. "The Difference Between Reacting and Responding." *Psychology Today*, October 5, 2021. https://www.psychologytoday.com/us/blog/the-power-prime/202110/the-difference-between-reacting-and-responding.

2 Lee, Emma. "What Is Neuroplasticity." Anahana, July 31, 2024. https://www.anahana.com/en/wellbeing-blog/physical-health/what-is-neuroplasticity.

3 "Neuroplasticity." *Psychology Today*. https://www.psychologytoday.com/us/basics/neuroplasticity.

4 Nerurkar, Aditi. *The 5 Resets: Rewire Your Brain and Body for Less Stress and More Resilience*. New York: HarperOne, 2024.

5 Dweck, Carol S. *Mindset: The New Psychology of Success.* New York: Random House, 2006.
6 "Growth Mindset." Massachusetts Institute of Technology. Teaching + Learning Lab. https://tll.mit.edu/teaching-resources/inclusive-classroom/growth-mindset/.

CHAPTER SIX: DON'T GET STUCK IN THE COMFORTABLE MISERY

1 "*Immunity to Change* by Robert Kegan and Lisa Lahey." Forge Coaching and Consulting, January 22, 2019. https://forgecoachingandconsulting.com/immunity-to-change-by-robert-kegan-and-lisa-lahey.
2 "Uncertainty Can Cause More Stress than Inevitable Pain." *ScienceDaily*, March 29, 2016. https://www.sciencedaily.com/releases/2016/03/160329101037.htm.
3 "Uncertainty Can Cause More Stress than Inevitable Pain."
4 "How to Work with Shame." National Institute for the Clinical Application of Behavioral Medicine, September 22, 2023. https://www.nicabm.com/program/a1-shame-fb2/.
5 Tedeschi, Richard G., Crystal L. Park, and Lawrence G. Calhoun. *Posttraumatic Growth: Positive Changes in the Aftermath of Crisis.* London: Routledge, 2014.
6 Tedeschi, et al. *Posttraumatic Growth.*
7 Collier, Lorna. "Growth After Trauma." American Psychological Association. *Monitor on Psychology* 47, no. 10 (November 2016). https://www.apa.org/monitor/2016/11/growth-trauma.
8 TEDx Talks. "Bouncing Back: An Experience with Post-traumatic

ENDNOTES

Growth Syndrome: Dave Sanderson: TEDxQueensU." YouTube, video, March 7, 2017. https://www.youtube.com/watch?v=LojjHV7FEJY.

9 Holmes, Thomas H., and Richard H. Rahe. "Holmes-Rahe Stress Inventory." American Institute of Stress, February 2024. https://www.stress.org/wp-content/uploads/2024/02/Holmes-Rahe-Stress-inventory.pdf.

10 Cohen, Sheldon, Michael L. Murphy, and Aric A. Prather. "Ten Surprising Facts About Stressful Life Events and Disease Risk." *Annual Review of Psychology* 70 (2019): 577–597.

11 Siegel, Dan. "Relationship Science and Being Human." Dr. Dan Siegel (website), December 17, 2013. https://drdansiegel.com/relationship-science-and-being-human/.

12 Kaufman, Scott Barry. "Post-traumatic Growth: Finding Meaning and Creativity in Adversity." *Scientific American*, April 20, 2020. https://blogs.scientificamerican.com/beautiful-minds/post-traumatic-growth-finding-meaning-and-creativity-in-adversity/.

13 Tsai, J., R. El-Gabalawy, W. H. Sledge, S. M. Southwick, and R. H. Pietrzak. "Post-traumatic Growth Among Veterans in the USA: Results from the National Health and Resilience in Veterans Study." *Psychological Medicine* 45, no. 1 (2015). https://doi.org/10.1017/s0033291714001202.

14 Tedeschi, Richard G. "Growth after Trauma." *Harvard Business Review*, July–August 2020. https://hbr.org/2020/07/growth-after-trauma.

15 Collier, Lorna. "Growth after Trauma." *Monitor on Psychology*, November 2016. https://www.apa.org/monitor/2016/11/growth-trauma.

ENDNOTES

CHAPTER SEVEN: SEEK SILVER LININGS

1 "Resilience." American Psychological Association. https://www.apa.org/topics/resilience.
2 Collingwood, Jane. "Realism and Optimism: Do You Need Both?" Psych Central, May 17, 2016. https://psychcentral.com/lib/realism-and-optimism-do-you-need-both.
3 Hu, Elise, and Andee Tagle. "The Power of Regret: How Examining Regret Can Help You Live a Meaningful Life." Life Kit. NPR, December 21, 2022. https://www.npr.org/2022/03/16/1087010308/the-power-of-regret-how-examining-regret-can-help-you-live-a-meaningful-life.
4 Pink, Daniel. "Summary of Our Mini-Survey on Regret." Daniel H. Pink (website), July 30, 2019. https://www.danpink.com/summary-of-our-mini-survey-on-regret/.
5 Steiner, Susie. "Top Five Regrets of the Dying." *The Guardian*, February 1, 2012. https://www.theguardian.com/lifeandstyle/2012/feb/01/top-five-regrets-of-the-dying.
6 Pink, Daniel H. *The Power of Regret*. Edinburgh, UK: Canongate, 2023.
7 TEDx Talks. "The Science of Regret: Marcel Zeelenberg: TEDxCollegeofEuropeNatolin." YouTube, video, September 27, 2016. https://www.youtube.com/watch?v=ZPCV3Oe1fYw.
8 Hayes, Adam. "Regret Avoidance: Meaning, Prevention, Market Crashes." *Investopedia*, September 29, 2022. https://www.investopedia.com/terms/r/regret-avoidance.asp.

ENDNOTES

CHAPTER EIGHT: FOCUS ON THE NEXT CLOSEST ALLIGATOR

1 Leonard, Jayne. "What to Know About Sensory Overload." *MedicalNewsToday*, Updated July 5, 2024. https://www.medicalnewstoday.com/articles/sensory-overload#symptoms.
2 Haden, Jeff. "How Emotionally Intelligent People Use the Yerkes-Dodson Law to Turn Stress into Optimal Performance and Achievement." *Inc.*, May 26, 2023. https://www.inc.com/jeff-haden/how-emotionally-intelligent-people-use-yerkes-dodson-law-to-turn-stress-into-optimal-performance-achievement.html.
3 Goleman, Dan. "The Sweet Spot for Achievement." *Psychology Today*, March 29, 2012. https://www.psychologytoday.com/us/blog/the-brain-and-emotional-intelligence/201203/the-sweet-spot-for-achievement.
4 Lopez-Garrido, Gabriel. "Bandura's Self-Efficacy Theory of Motivation in Psychology." Simply Psychology, July 10, 2023. https://www.simplypsychology.org/self-efficacy.html.
5 Lopez-Garrido. "Self-Efficacy."
6 D'Alessandro, Jon. "How to Develop Self-Efficacy: Enactive Mastery Experiences." High Potential Coaching, January 7, 2024. https://www.bringambition.com/post/enactive-mastery-experiences.
7 Schachter, Stanley, Norris Ellertson, Dorothy McBride, and Doris Gregory. "An Experimental Study of Cohesiveness and Productivity." *Human Relations* 4, no. 3 (1951): 229–38. https://doi.org/10.1177/001872675100400303.

8 Berg, Sara. "What Doctors Wish Patients Knew About Decision Fatigue." American Medical Association, November 19, 2021. https://www.ama-assn.org/delivering-care/public-health/what-doctors-wish-patients-knew-about-decision-fatigue.
9 "Stress and Decision-Making During the Pandemic." American Psychological Association, October 26, 2021. https://www.apa.org/news/press/releases/stress/2021/october-decision-making.
10 "Stress and Decision-Making During the Pandemic."
11 Clear, James. "How Willpower Works: How to Avoid Bad Decisions." James Clear (website), July 20, 2018. https://jamesclear.com/willpower-decision-fatigue.
12 Milyavsky, Maxim, David Webber, Jessica Fernandez, Arie W. Kruglanski, Amit Goldenberg, Gaurav Suri, and James Gross. "To Reappraise or Not to Reappraise? Emotion Regulation Choice and Cognitive Energetics." *Emotion* 19 (2018). Preprint available at https://doi.org/10.31234/osf.io/j4ak5.
13 "Cognitive Reappraisal Strategy for Emotional Regulation." Cognitive Behavioral Therapy Los Angeles, October 9, 2023. https://cogbtherapy.com/cbt-blog/2014/5/4/hhy104os08dekc537dlw7nvopzyi44.

CHAPTER NINE: YOU'RE BETTER LUCKY THAN GOOD

1 Schoemaker, Paul J. H. "Forget Dumb Luck—Try Smart Luck: Strategies to Get Lady Fortune on Your Side." *Management and Business Review* 1, no. 2 (2021).
2 Kashdan, Todd B., David J. Disabato, Fallon R. Goodman, and Carl Naughton. "The Five Dimensions of Curiosity." *Harvard Business Review*, September–October 2018. https://hbr.org/2018/09/the-five-dimensions-of-curiosity.

ENDNOTES

3 Kashdan, et al. "The Five Dimensions of Curiosity."

4 Insinna, Valerie. "To Get More Female Pilots, the Air Force Is Changing the Way It Designs Weapons." *Air Force Times*, August 19, 2020. https://www.airforcetimes.com/news/your-air-force/2020/08/19/to-get-more-female-pilots-the-air-force-is-changing-the-way-it-designs-weapons/.

5 Novelly, Thomas. "Female Air Force Pilots Would Be Able to Safely Pee In-Flight During Long Missions with New Tech Being Tested." *Military.com*, March 20, 2023. https://www.military.com/daily-news/2023/03/20/female-air-force-pilots-would-be-able-safely-pee-flight-during-long-missions-new-tech-being-tested.html.

6 Holiday, Ryan. "What Is Luck and What Is Not." Daily Stoic, October 3, 2019. https://dailystoic.com/what-is-luck-and-what-is-not/.

7 Thibodeaux, Wanda. "Here's What Practicing Does to Your Brain (and How to Do It Right)." *Inc.*, March 13, 2017. https://www.inc.com/wanda-thibodeaux/heres-what-practicing-does-to-your-brain-and-how-to-do-it-right.html.

8 Pluchino, Alessandro, Alessio Emanuele Biondo, and Andrea Rapisarda. "Talent Versus Luck: The Role of Randomness in Success and Failure." *Advances in Complex Systems* 21, no. 3–4 (2018). https://doi.org/10.1142/s0219525918500145.

9 "Serendipity (n.)." *Online Etymology Dictionary*. https://www.etymonline.com/word/serendipity.

10 de Rond, Mark. "The Structure of Serendipity." Working Paper No. 05/07, Cambridge Judge Business School, University of Cambridge, 2005. https://www.jbs.cam.ac.uk/wp-content/uploads/2020/08/wp0507.pdf.

11 American Psychological Association. "Transforming Uncertainty

into Opportunity: The Serendipity Mindset." YouTube, video, May 14, 2020. https://www.youtube.com/watch?v=RykaAg7eT3s&t=21s.
12 "Percy Spencer: Microwave Inventor." Smithsonian National Museum of American History. Lemelson Center for the Study of Invention and Innovation, May 12, 2014. Accessed January 11, 2024. https://invention.si.edu/node/1145/p/431-percy-spencer-microwave-inventor.
13 "The Serendipity Test." *Nature*, January 31, 2018. https://www.nature.com/articles/d41586-018-01405-7.
14 "The Serendipity Test."
15 Wiseman, Richard. "The Luck Factor." *The Skeptical Inquirer*, May–June 2003.

Chapter Ten: Avoid Get Home–itis

1 Hoomans, Joel. "35,000 Decisions: The Great Choices of Strategic Leaders." Roberts Wesleyan College. The Leading Edge, March 20, 2015. https://go.roberts.edu/leadingedge/the-great-choices-of-strategic-leaders.
2 Einhorn, Cheryl Strauss. "How to Make Rational Decisions in the Face of Uncertainty." *Harvard Business Review*, August 28, 2020. https://hbr.org/2020/08/how-to-make-rational-decisions-in-the-face-of-uncertainty.
3 TED. "How Great Leaders Inspire Action: Simon Sinek: TED." YouTube, video, May 4, 2010. https://www.youtube.com/watch?v=qp0HIF3SfI4.
4 "Vision: It's Our Promise to Protect." U.S. Air Force. https://www.airforce.com/vision.

ENDNOTES

5 "57th Wing." Nellis Air Force Base. https://www.nellis.af.mil/Units/57-WG/.

6 Rhodes, Kay. "Beyonce Says: So Long, Sasha Fierce!" *Hollywood Gossip*, January 20, 2010. https://www.thehollywoodgossip.com/2010/01/beyonce-says-so-long-sasha-fierce/.

CHAPTER ELEVEN: CHECK YOUR SIX

1 NDTV. "Videos: 2 Fall off Plane, Some Huddled on Aircraft Wing in Kabul Mayhem." YouTube, video, August 16, 2021. https://www.youtube.com/watch?v=mIQPSgnG3T0.

2 "US to Conduct New Interviews into the Deadly 2021 Bombing at Kabul Airport." Al Jazeera, September 15, 2023. https://www.aljazeera.com/news/2023/9/15/us-to-conduct-new-interviews-into-the-deadly-2021-bombing-at-kabul-airport.

3 Weissbourd, Richard, Milena Batanova, Virginia Lovison, and Eric Torres. "Loneliness in America: How the Pandemic Has Deepened an Epidemic of Loneliness and What We Can Do About It." Harvard Graduate School of Education, 2021.

4 Weissbourd, et al. "Loneliness in America."

5 U.S. Department of Health and Human Services. "New Surgeon General Advisory Raises Alarm About the Devastating Impact of the Epidemic of Loneliness and Isolation in the United States." Press release, May 3, 2023. https://www.hhs.gov/about/news/2023/05/03/new-surgeon-general-advisory-raises-alarm-about-devastating-impact-epidemic-loneliness-isolation-united-states.html.

6 "Why Americans Are Lonelier and Its Effects on Our Health." *PBS News Weekend*, PBS, January 8, 2023. https://www.pbs

.org/newshour/show/why-americans-are-lonelier-and-its-effects-on-our-health.

7 Cox, Daniel A. "The State of American Friendship: Change, Challenges, and Loss." The Survey Center on American Life, June 8, 2021. https://www.americansurveycenter.org/research/the-state-of-american-friendship-change-challenges-and-loss/.

8 Cost, Ben. "Americans Have Fewer Friends Than Ever Before: Study." *New York Post*, July 27, 2021. https://nypost.com/2021/07/27/americans-have-fewer-friends-than-ever-before-study/.

9 Swan, Gary E., and Dorit Carmelli. "Curiosity and Mortality in Aging Adults: A 5-Year Follow-Up of the Western Collaborative Group Study." *Psychology and Aging* 11, no. 3 (1996). https://doi.org/10.1037//0882-7974.11.3.449.

10 Sakaki, Michiko, Ayano Yagi, and Kou Murayama. "Curiosity in Old Age: A Possible Key to Achieving Adaptive Aging." *Neuroscience & Biobehavioral Reviews* 88 (May 2018). https://doi.org/10.1016/j.neubiorev.2018.03.007.

11 Jones, Daniel. "The 36 Questions That Lead to Love." *New York Times*, Modern Love, January 9, 2015. https://www.nytimes.com/2015/01/09/style/no-37-big-wedding-or-small.html.

12 Aron, Arthur, Edward Melinat, Elaine N. Aron, Robert Darrin Vallone, and Renee J. Bator. "The Experimental Generation of Interpersonal Closeness: A Procedure and Some Preliminary Findings." *Personality and Social Psychology Bulletin* 23, no. 4 (April 1997). https://doi.org/10.1177/0146167297234003.

13 Holt-Lunstad, Julianne. "Social Connection as a Public Health Issue: The Evidence and a Systemic Framework for Prioritizing the 'Social' in Social Determinants of Health."

ENDNOTES

Annual Review of Public Health 43, no. 1 (2022). https://doi.org/10.1146/annurev-publhealth-052020-110732.
14 Holt-Lunstad. "Social Connection."
15 "Social Networks and Health: Communicable but Not Infectious." Harvard Medical School. Harvard Health Publishing, December 1, 2011. https://www.health.harvard.edu/staying-healthy/social-networks-and-health-communicable-but-not-infectious.
16 Theisen, Angela. "Is Having a Sense of Belonging Important?" Mayo Clinic Health System, December 8, 2021. https://www.mayoclinichealthsystem.org/hometown-health/speaking-of-health/is-having-a-sense-of-belonging-important.
17 "The Importance of Connections on Our Well-Being." Berkeley Executive Education. https://executive.berkeley.edu/thought-leadership/blog/importance-connections-our-well-being.

CHAPTER TWELVE: YOU'RE SAYING THERE'S A CHANCE

1 "Inspire." *Collins English Dictionary*. https://www.collinsdictionary.com/dictionary/english/inspire.
2 Nelson, McKenzie. "Thunderbirds' Female Lead Solo Pilot to Fly for KC Air Show." KSHB 41 Kansas City News, July 1, 2021. https://www.kshb.com/news/local-news/thunderbirds-sole-female-pilot-front-and-center-literally-for-kc-air-show.

ABOUT THE AUTHOR

Michelle "MACE" Curran is a former United States Air Force fighter pilot with nearly two thousand hours of F-16 flying time. She flew combat missions in Afghanistan and honed her skills across the globe, becoming the second woman in history to serve as the Lead Solo Pilot for the Thunderbirds, the Air Force's elite demonstration team. Known for her signature upside-down maneuvers, Mace performed for millions, inspiring audiences at air shows and with flyovers for the Super Bowl, the Daytona 500, and the Indy 500. Today, she channels her unique perspective into relatable storytelling to empower others to make bold choices and find their inner courage.

RAISING READERS
Books Build Bright Futures

Thank you for reading this book and for being a reader of books in general. As an author, I am so grateful to share being part of a community of readers with you, and I hope you will join me in passing our love of books on to the next generation of readers.

Did you know that reading for enjoyment is the single biggest predictor of a child's future happiness and success?

More than family circumstances, parents' educational background, or income, reading impacts a child's future academic performance, emotional well-being, communication skills, economic security, ambition, and happiness.

Studies show that kids reading for enjoyment in the US is in rapid decline:

- In 2012, 53% of 9-year-olds read almost every day. Just 10 years later, in 2022, the number had fallen to 39%.
- In 2012, 27% of 13-year-olds read for fun daily. By 2023, that number was just 14%.

Together, we can commit to **Raising Readers** and change this trend. How?

- Read to children in your life daily.
- Model reading as a fun activity.
- Reduce screen time.
- Start a family, school, or community book club.
- Visit bookstores and libraries regularly.
- Listen to audiobooks.
- Read the book before you see the movie.
- Encourage your child to read aloud to a pet or stuffed animal.
- Give books as gifts.
- Donate books to families and communities in need.

Books build bright futures, and **Raising Readers** is our shared responsibility.

For more information, visit **JoinRaisingReaders.com**

Sources: National Endowment for the Arts, National Assessment of Educational Progress, WorldBookDay.org, Nielsen BookData's 2023 "Understanding the Children's Book Consumer"